Gastroenterology Research and Clinical Developments

nova
Medicine & Health
New York

Gastroenterology Research and Clinical Developments

Probiotics and their Role in Health and Disease
Yusef Olsen (Editor)
2021. ISBN: 978-1-53619-965-9 (Hardcover)
2021. ISBN: 978-1-53619-997-0 (eBook)

More information about this series can be found at
https://novapublishers.com/product-category/series/gastroenterology-research-and-clinical-developments/

Amir M. Mortazavian
Zahra Yari
Nasim Khorshidian

Effects of Healthful Bioactive Compounds on the Gastrointestinal Tract

nova
Medicine & Health
New York

DOI: https://doi.org/10.52305/YUYD5491

NOTICE TO THE READER

Additional color graphics may be available in the e-book version of this book.

Library of Congress Cataloging-in-Publication Data

ISBN: 978-1-68507-624-5

Published by Nova Science Publishers, Inc. † New York

Contents

Preface

Optimal performance of the liver and gastrointestinal tract is vital for general human health. The gastrointestinal system is the main site of nutrient digestion and absorption, and therefore has substantial effects on providing energy and growth for the body. All nutrient and non-nutrient components have an impact on the gastrointestinal system, either locally or systemically. Undeniably, gastrointestinal disorders, with increasing lifetime prevalence, have a profound effect on human health and reduce the quality of life.

Nowadays, therapeutic options are limited to medical therapy, including pharmacologic approaches and surgical intervention or lifestyle changes aimed at dietary modifications and increased physical activity. Certainly, dietary modifications can be effective in preventing, treating, and alleviating associated symptoms and complications. In addition, to improve whole dietary patterns, the components of foods, collectively or independently, also have a substantial role in their therapeutic features. These components, particularly bioactive compounds, confer unique health benefits and aid in management of various diseases.

Recently, investigators have paid much attention to chemopreventive activity of bioactive dietary compounds in the treatment and prevention of gastrointestinal diseases. Fruits, vegetables and edible plants are the main source of biologically active compounds with anti-inflammatory, antioxidant, and protective properties. Most importantly, no major side effects have been reported for these compounds until now. Therefore, this book offers novel targets for prevention, treatment and management of liver and gastrointestinal diseases by providing evidence of effectiveness of bioactive compounds of functional foods. Furthermore, intracellular and molecular pharmacological mechanisms of action of bioactive compounds will also be discussed.

It seems that bioactive compounds bring new hope for patients with gastrointestinal disorders through their anti-inflammatory and anti-oxidative potential.

The authors hope that the book will be a useful reference for researchers, product developers, health professionals and students.

Amir M. Mortazavian
Zahra Yari
Nasim Khorshidian

Chapter 1

An Overview of the Gastrointestinal Tract and Its Importance in Human Health

Abstract

The gastrointestinal (GI) tract is a complex system passing from the mouth to the anus. Food taken in through the mouth is digested, nutrients and energy are absorbed and the waste is expelled at the anus. The stomach and small intestine are principally involved in digestion and absorption that is carried out through physical and chemical mechanisms. The large intestine is responsible for dehydration and compression of waste before its elimination. It is considered a neuroendocrine system and therefore, the dynamic interplay of multiple hormonal, humoral, and neural pathways is crucial for the proper function of intestinal motility, secretion, absorption, and defense against pathogens. GI tract harbors a community of microorganisms known as gut microbiota which has a substantial role in the pathogenesis of various diseases. This chapter describes an overview of different segments of the GI tract and various diseases associated with these parts and the involved mechanisms.

Keywords: disease, gastrointestinal tract, health, microbiota

1. Introduction

The gastrointestinal tract (GIT) is a complex system responsible for digestion, absorption of nutrients, and disposal of waste substances and indigestible remnants (Johnson et al. 2012). The digestive tract is considered as a neuroendocrine system. Therefore, the dynamic interplay of multiple hormonal, humoral, and neural pathways is crucial for the proper function of intestinal motility, secretion, absorption, and defense against pathogens (Schubert 2008; Mayer 2011). Approximately 70% of the immune cells are located in the GIT (termed gut-associated lymphoid tissue (GALT)), playing an essential role in the immune system. Both the autonomic and enteric nervous systems are involved in regulating digestion processes (Mason et al.

2008). Moreover, there is a region specificity and complexity in gastrointestinal digestion (Barrett, Malley, and Naglieri 2006; Farre and Tack 2013). The digestive system expends more than half of the body's energy production devoted to food digestion (Watson and Preedy 2012).

Digestion involves a series of interrelated metabolic processes that turn ingested carbohydrates, proteins, fats, and other nutrients into a form that the body can utilize for nourishment, growth, maintenance and repair, depending on the requirements and conditions. Digestion initiates through seeing, tasting, and smelling of food involved in signaling to the brain and then stimulating digestive juices secretion. The first step is releasing saliva, which, in addition to facilitating swallowing, contains digestive enzymes, mostly amylase, that play an important role in the early stages of carbohydrate hydrolysis. Then, the food bolus is delivered through the esophagus to the stomach by a peristaltic mechanism (Woda et al. 2010). Digestion begins with lowering the pH in the stomach and food is crushed and mixed with mechanical pressure. Chyme is a mixture produced by breaking up and acidifying the stomach contents. In addition to mechanical actions, digestive enzymes like pepsin and lipase also help digestion in the stomach. This multi-directional movement exposes food molecules to enzymes and hydrochloric acid (HCl), breaking down macronutrients into their components. The predigested chyme resulting from mechanical and enzymatic degradations is transported to the small intestine. Gastric emptying rate is a key parameter of digestion mainly influenced by food composition, structure, and biological factors (Golding and Wooster 2010; Ballance et al. 2013).

The small intestine is divided into three successive segments. The anterior part is named the duodenum, entry site of digestive secretions from the pancreas and liver. The other two parts are called jejunum and ileum, respectively (Barrett et al., 2006). At this stage, sodium bicarbonate ($NaHCO_3$), to some extent, neutralizes the acidic pH of the chyme so that the pancreatic enzymes can be effective. Chyme must be sufficiently acidified to stimulate the secretion of bicarbonate and pancreatic enzymes; therefore, adequate gastric acid is crucial for proper digestion. Pancreatic secretions are a complex mixture of proteases, amylases, and lipases that perform the major digestive process. In addition to lipase, bile also helps digest fats in the intestine. Bile is produced by the liver and facilitates pancreatic lipase activity by emulsifying dietary fats. Most bile salts are eventually reabsorbed in the ileum and enter into the enterohepatic cycle for reuse (Barrett, Malley, and Naglieri 2006).

Non-absorbed and non-digested materials move down to the large intestine. In addition to the formation, storage, and elimination of feces, the large intestine also has the function of water and electrolytes absorption. The gastrointestinal tract harbors distinct microbial populations. It is well defined that microbiota composition influences the incidence of disease and general health (He, Magnuson, and Giusti 2005; Booijink et al. 2010; van den Bogert et al. 2011).

2. Main Disorders of the Gastrointestinal Tract and the Involved Mechanisms

According to the National Institutes of Health, more than seventy million Americans suffer from digestive disorders, 15% of them daily. About one hundred million of these cases have been reported to experience gastroesophageal reflux disease at least once a month. In contrast, many others suffer from peptic ulcers, irritable bowel syndrome and constipation. The incidence of gastrointestinal disorders is rising, but the current outbreak appears to be under-estimated. Self-treatment is a common practice among people with gastrointestinal disorders, especially stomach problems and they essentially try herbal therapy, including ginger, peppermint, licorice, aloe vera, etc. Today, great attention is paid to prebiotic and probiotic products in digestive health (Wolf 2012; Brandon 2004).

2.1. Peptic Ulcer Disease

Peptic ulcer disease (PUD) is a severe and multifactorial disease encompassing gastric and duodenal ulcers (Valle 2005). For over a century, peptic ulcer disease has been one of the most prevalent diseases globally, with high morbidity and mortality rates (Sung et al. 2010). Although PUD can occur at all ages, it is more prevalent between 30 and 60 (Martins et al. 2002). The prevalence differs between the duodenal and gastric ulcers. Still, in general, it affects approximately 4 to 8% of the global population, and nearly 10% of people experience PUD in their lifetime and it is more common in men (Aro et al. 2006; Sonnenberg and Everhart 1996; Bandyopadhyay et al. 2001; Calam and Baron).

Peptic ulcer develops with a chronic trend, characterized by an imbalance between offensive (hydrochloric acid, pepsin, and *Helicobacter pylori*) and defensive factors (mucin, prostaglandin, bicarbonate, nitric oxide, and growth factors) (Willemijntje and Pankaj 2001), resulting in a mucosal lining lesion of the upper digestive tract. In this respect, major therapeutic approaches focus on reducing acid secretion or enhancing protective factors (Willemijntje and Pankaj 2001; Valle 2005), but still, the prevention or treatment of peptic ulcers is challenging. Factors that increase acid secretion or reduce mucus and bicarbonate secretion can contribute to the disease, such as alcohol consumption, smoking, *H. pylori* infection, and non-steroidal anti-inflammatory drugs (NSAIDs). The role of genetics is also highlighted in PUD, as it has been shown that the risk of peptic ulcer is three times higher in children whose parents have ulcers (Toneto, Oliveira, and Lopes 2011; de Lira Mota et al. 2009). However, some PUD patients exhibit no evident cause (Jones 2006). Since there is a strong relation between PUD and emotional stress and obesity, following a healthy diet and lifestyle modification may play a role in the onset and course of PUD (Everhart, Bryd-Holt, and Sonnenberg 1998; Kalichman, Livshits, and Kobyliansky 2006; Kurata and Nogawa 1997). Accordingly, the main purpose of diet therapy is to repair and protect the gastrointestinal lining and relieve pain, which leads to better digestion and contributes to a good nutritional status (Reis 2003).

Although various drugs are under investigation to suppress offensive agents in peptic ulcer, there is still no definite treatment (Farzaei, Abdollahi, and Rahimi 2015). It seems that some herbal medicine can deal with offensive factors. These substances can be applied in treatment because of being relatively less expensive with better patient tolerance if their safety and efficacy are confirmed (Goel and Sairam 2002).

2.2. Gastroesophageal Reflux Disease

Gastroesophageal reflux disease (GERD) is defined by two dominant symptoms, including heartburn or acid regurgitation, with a weekly frequency (El-Serag et al. 2013). The prevalence in adults is increasing and varies widely from 10% to 30% due to the difference in nations. The incidence is estimated at 5/1000 person-years in Western populations (El-Serag et al. 2013). It has been reported that approximately 20% of Americans experience GERD-related symptoms at least once a week (Klauser, Schindlbeck, and Müller-Lissner 1990; Locke et al. 1997), which is more prevalent amongst obese individuals (Suter et al. 2004). Over time, repetitive exposure of the esophagus to gastric acid causes increased oxidative stress, increasing the risk of Barrett's esophagus or distal adenocarcinoma of the esophagus, two severe complications of chronic GERD (Romero et al. 1997; Trudgill, Kapur, and Riley 1999). In addition, GERD is associated with a reduction in health-related quality of life and labor productivity (Ronkainen et al. 2006; Wahlqvist, Reilly, and Barkun 2006). Diagnostic tests and treatment approach also impose high costs on individuals and the community (Mason and Hungin 2005). No satisfactory physiological documents explaining the etiology of symptoms or a therapeutic target are yet achieved.

2.3. Constipation

Problems with bowel movements are a frequent complaint leading to approximately 20% of gastrological outpatient clinical visits. According to the criteria used for defining constipation, the lifetime prevalence of constipation can be very variable with estimates between 2% and 34% (Sandler and Drossman 1987; Whitehead et al. 1991; Drossman et al. 1993; Pare et al. 2001). Gut transit refers to the time taken for intraluminal contents to traverse the gastrointestinal tract which is the functional consequence of tonic and phasic gut contractions (Kellow et al. 2006). Average whole-gut transit time is estimated to be 30-40 h, with a wide variation between individuals (Kim and Rhee 2012).

Gastrointestinal microbiota, enteric nervous system, and central nervous system are major regulators of gut motility; disturbance in any of these interrelated systems may contribute to symptoms of constipation (Dimidi et al. 2017). In most cases, constipation is followed by secondary causes such as immobility, low dietary fiber, and dehydration. Other factors, including retention of fecal mass and imbalance of intestinal microflora, also contribute to constipation. Female gender is found to be a risk factor and physical exercise is identified to be a protective factor. Bowel perforation, hemorrhoids, anal fissures and rectal prolapse are complications that are attributed to chronic constipation (Battistella and Alibhai 2004). Additionally, increased risk of constipation-related psychological morbidity has been observed in women (Mason, Serrano-Ikkos, and Kamm 2000).

2.4. Diarrhea

Diarrhea is a gut motility disorder associated with changes in fecal volume, fluidity, and frequency of defecation (Hall 2009). Diarrhea is an intestinal infection caused by viruses or bacteria and is considered a significant clinical problem (Casburn-Jones and Farthing 2004; Kosek, Bern, and Guerrant 2003). Diarrhea is one of the leading causes of morbidity and mortality in the elderly in developed countries (Taylor and Echeverria 1993). It is the second leading cause of death among children under the age of 5 (Liu et al. 2012).

Diarrhea is also a hallmark manifestation in several gastrointestinal diseases, including diarrhea-predominant irritable bowel syndrome (IBS) and inflammatory bowel disease (IBD) (Excellence 2008; Hanif Palla and Gilani 2015). Hence, diarrhea and morbidity will also cause sickness absenteeism, social inconvenience and higher medical expenses (Camilleri and Murray 2008).

The prevalence of chronic diarrhea is 2–7%, but higher in children and the elderly (Peppas et al. 2008). Among the viruses that cause diarrhea, rotavirus and calicivirus are predominate (Mrukowicz et al. 1999). Rotaviral diarrhea is acute and may cause lactose intolerance by damaging mature enterocytes (Ciarlet and Estes 2001; Isolauri et al. 1994). In bacterial diarrhea cases, those caused by *Campylobacter jejuni*, *Salmonella*, *Shigella,* and enterotoxigenic *Escherichia coli* are more common.

Prolonged use of antibiotics may predispose people to diarrhea. Typically, antibiotic-dependent diarrhea is mild except for those caused by *Clostridium difficile* infection, which may result in pseudomembranous colitis (Kyne and Kelly 2001; Barbut et al. 2000). Maintenance of water and electrolytes balance by oral rehydration solution (ORS) is the most common therapy in diarrhea. ORS contains sodium, potassium, chloride, citrate, and glucose (Atia and Buchman 2009; Munos, Walker, and Black 2010). Diet therapy for diarrhea includes administration of zinc gluconate, dietary soluble fiber and probiotic products, albeit these recommendations have contradictory results (Gregorio et al. 2007; Basu et al. 2007; Narayanappa 2008; Passariello et al. 2011; Hoekstra et al. 2004).

2.5. Irritable Bowel Syndrome

Irritable bowel syndrome is a functional disorder of the gastrointestinal system. According to the National Institutes of Health, IBS affects approximately 20% of the American adult population and tends more likely to be observed in women and young people. In one-half of cases, the initial onset of symptoms occurs before the age of 35 (NDDIK 2010).

There is a wide range of symptoms associated with IBS, varying from cramping, abdominal pain, tenderness and bloating to severe constipation and diarrhea. Patients are divided into three subgroups in terms of symptoms: diarrhea-predominant, constipation-predominant, or mixed subtype (Kassinen et al. 2007).

The definitive cause and pathophysiology of IBS are unknown, although an alteration in the interplay between the nervous system and brain-gut axis seems to be the main causes. Stress and emotional distress can exacerbate symptoms (Parkes et al. 2008). In addition, evidence has mentioned various factors, including gut motility disturbance, neurotransmitters imbalance, visceral hypersensitivity, infection and inflammation of GIT. Although IBS is not life-threatening or is not the cause of other serious diseases, it can significantly reduce the quality of life (Kassinen et al. 2007; Horwitz and Fisher 2001). Recently, instability of gut microbiota (dysbiosis) has been introduced as a modifiable risk factor in the pathogenesis of IBS (King, Elia, and Hunter 1998; Nobaek et al. 2000; Putignani et al. 2015).

2.6. Inflammatory Bowel Disease

Inflammatory bowel disease is an idiopathic intestinal chronic immune-mediated disease characterized by remission and relapse course (Podolsky 2002). IBD typically has its peak of onset during young adulthood. The incidence and prevalence of IBD are increasing worldwide, particularly in developed countries that have been estimated as high as 1% (Kaulmann and Bohn 2016).

The major clinical manifestations of IBD include gut bleeding, obstruction, abdominal pain and diarrhea caused by extensive intestinal structural damage, leading to the disruption and ulcer formation in the epithelial barrier (da Silva et al. 2010; Biasi et al. 2011). Systemic symptoms related to IBD generally include fever, weight loss, wasting, and an ultimately increased risk of cancer (Larsen, Bendtzen, and Nielsen 2010; Kaulmann and Bohn 2016). In addition, more than a quarter of the patients experience extra-intestinal complications, including peripheral arthritis, erythema nodosum, pyoderma gangrenosum, episcleritis, anemia, etc. (Biasi et al. 2011; Larsen, Bendtzen, and Nielsen 2010). Inflammatory bowel disease encompasses two chronic and distinct clinical entities: Crohn's disease (CD) and ulcerative colitis (UC), characterized by an exaggerated immune response to gut microbiota leading to impaired epithelial barrier function. Although the two diseases mentioned above are distinct diseases or two forms of IBD, they have not been clarified yet (Drewes et al. 2006; Shih and Targan 2008; Söderholm et al. 2004). The prevalence of Crohn's disease and ulcerative colitis has been estimated at approximately 200 cases per 100000 population in the United States. In most cases, ileum and colon have been involved (Watson and Preedy 2012). Similar to CD, UC is characterized by mucosal barrier ulceration in the colon and rectum (Watson and Preedy 2012).

Despite numerous studies, the etiology and pathogenesis of IBD have not been clarified yet. However, complex factors, such as genetic predisposition, enteric immune system and environmental factors, and even gut microbiome imbalance, are involved in the pathogenesis of this disease (Hold et al. 2014; Baumgart and Carding 2007; Podolsky 2002). According to epidemiological studies, diet is the most crucial element among environmental factors, which can be an imperative target in treating IBD (Hou, Abraham, and El-Serag 2011; Lee et al. 2015).

Patients with CD and UC often depend on lifelong medication. Varieties of medical and complementary therapies are provided for IBD patients' treatment. They are usually treated with aminosalicylates, corticosteroids, immunosuppressive agents, anti-tumor necrosis factor (TNF) antibodies and antibiotics (Bernstein et al. 2012; Rocchi et al. 2012; Watson and Preedy 2012). Some of these agents may increase the risk of infection and malignancy (Lichtenstein et al. 2012; Toruner et al. 2008). Moreover, some patients are resistant to pharmacotherapy; thus, colectomy might be required to control inflammation (Lee et al. 2015). Nutritional recommendations, along with conventional therapies, can improve the symptoms of illness and quality of life in patients with IBD (Brown, Rampertab, and Mullin 2011). On the other hand, many patients are increasingly turning to complementary and alternative medicine therapies (Fernández et al. 2012; Lakatos et al. 2010). Many studies have been conducted on anti-inflammatory agents (Chu 2014; Gupta et al. 2014). Although the protective and therapeutic effects of many polyphenols have been noted in IBD, the underlying mechanisms of their anti-inflammatory properties remain primarily speculative. On the other hand, achieving a therapeutic dose in a typical diet is almost far from the mind (Romier et al. 2009).

2.7. Celiac Disease

Celiac disease is defined as an autoimmune-mediated enteropathy characterized by impairment of the mucosal barrier, villous atrophy, and subsequently, malabsorption of nutrients, precipitated by exposure to dietary gluten (Fasano and Catassi 2001; Green, Lebwohl, and Greywoode 2015; Assimakopoulos, Papageorgiou, and Charonis 2011) (Figure 1.1). Gluten is a storage protein found in wheat, rye, and barley (Iwańczak and Iwańczak 2012) and this permanent intolerance occurs in genetically predisposed individuals (Rubio-Tapia et al. 2013). Concerning etiology, infections and imbalance of gut microbiota have also been mentioned (Schuppan, Junker, and Barisani 2009).

The worldwide prevalence of the celiac disease is on the rise and is estimated to be around 1% in Western populations (Riddle, Murray, and Porter 2012; Dubé et al. 2005). Celiac disease is diagnosed through anti–tTG-IgA (tissue transglutaminase) assessment and histological examination of the biopsy specimens of the small intestine (Rubio-Tapia et al. 2013). At present, the only effective treatment for celiac disease is a lifelong dietary exclusion of gluten-containing grains (Green, Lebwohl, and Greywoode 2015), which requires patient education by a dietitian, motivation, and patient follow-up (See and Murray 2006). Even minor amounts of gluten (50 mg/day) can lead to relapse; therefore, complete elimination of gluten and its derivatives is required (Catassi et al. 2007). Dietary restrictions in celiac patients lead to loss of quality of life and produce psychological, emotional and economic stress (Lee and Newman 2003). Cost, poor palatability, limited availability of gluten-free products and insufficient labeling are barriers to dietary compliance (Stevens and Rashid 2008). Therefore, there is a need to develop therapeutic approaches other than the gluten-free diet for patients with celiac disease (Tennyson et al. 2013).

2.8. Non-Alcoholic Fatty Liver Disease

Non-alcoholic fatty liver disease (NAFLD) is defined as a common liver disorder that encompasses a broad spectrum of liver pathologies, including simple hepatic steatosis (determined as hepatic TG >5% by liver weight) to inflammatory non-alcoholic steatohepatitis (NASH), fibrosis, subsequently progressed to cirrhosis and ultimately becoming a leading cause of the end-

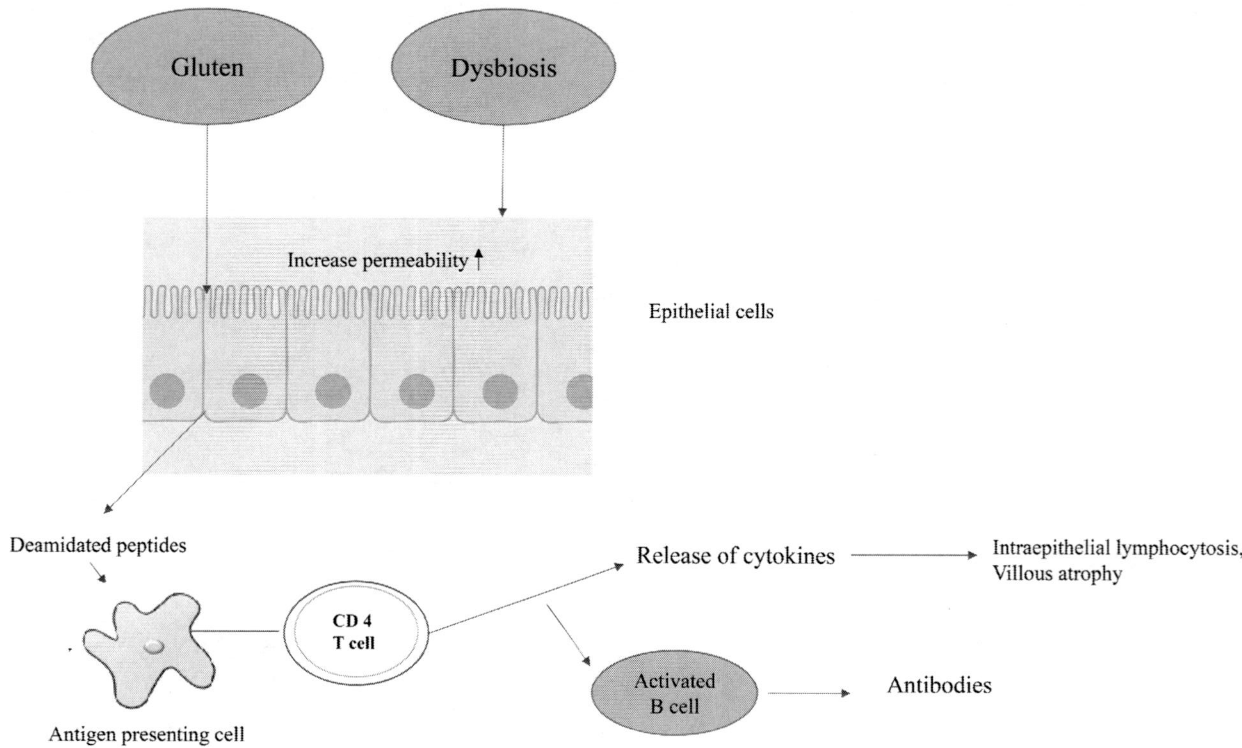

Figure 1.1. Schematic presentation of celiac disease pathogenesis.

stage liver failure or hepatocellular carcinoma (Angulo 2002; Michelotti, Machado, and Diehl 2013; Lazo and Clark 2008).

Non-alcoholic fatty liver disease describes a condition caused by abnormal hepatic triglyceride (TG) accumulation in the absence of excessive alcohol consumption (Angulo 2002). The pathogenesis of NASH is multifactorial. Interaction between insulin resistance, inflammation, oxidative stress, metabolic disorders and mitochondrial dysfunction is involved in the pathogenicity (Kumar et al. 2012; Ahmed, Barakat, and Almobarak 2012; Ma and Li 2006). The imbalance between free fatty acid acquisition and inefficient β-oxidation, esterification, or both are the major causes of hepatocytes lipid droplet accumulation, the major hallmark of NAFLD disposal (Musso, Gambino, and Cassader 2009; Charytoniuk et al. 2017).

NAFLD has a relatively benign course in most patients and remains steady for years (Angulo 2002); therefore, the outbreak cannot be well-estimated. In addition, it depends on ethnicity, body weight and the presence of associated diseases such as diabetes mellitus or dyslipidemia (Lazo et al. 2013; Caballería et al. 2012). The universal prevalence of NAFLD in the adult population is estimated to be 10–24%, with the highest rate in industrialized communities (Chalasani et al. 2012; LaBrecque et al. 2014). This percentage increases to over 50% with obesity and 70% in type II diabetic patients (Ong and Younossi 2007; Browning et al. 2004). Although body mass index (BMI), particularly central obesity is strictly associated with NAFLD, recent studies indicated that 20–40% of patients had normal body weight (Yasutake et al. 2009; Nakamuta et al. 2008; Maximos et al. 2015). Increasing the prevalence of NAFLD is not limited to adults, but also occurs in children (3–10%) and is growing up to 40–70% among obese children (Bellentani et al. 2010). NAFLD has no specific symptom to be distinguished, thus, liver biopsy is the gold standard for both clinical diagnosis and staging (Neuschwander-Tetri et al. 2010). Although NAFLD is benign, it has been reported that in less than two decades, near 40% of NAFLD cases progress to fibrosis and 50% develop NASH (Argo et al. 2009; Ekstedt et al. 2006). Several predisposing factors are involved in the development and progression of NAFLD. Despite progress in understanding the procedure and mechanisms, up to now, there is no effective proven treatment for NAFLD (Pan et al. 2014).

Nowadays, therapeutic options are limited to lifestyle changes and dietary modifications aimed at weight loss, increased physical activity, and few pharmacological therapies (Sahebkar 2011; Hajiaghamohammadi, Ziaee, and Samimi 2012; Dyson, Anstee, and McPherson 2013). Besides the importance of improving whole diet intake, foods' components also have a significant role in its therapeutic features (Kani et al. 2017). Currently, numerous functional food components are under investigation for their potential role in the prevention and treatment of diseases and promotion of health (Clare M Hasler 1998).

The exciting and rapidly growing trend of functional foods constructs noteworthy attractive fields and opportunities for public health; however, it has not yet come up with a legal meaning in the United States (Hasler and Brown 2009). Functional foods have been defined as foods that, due to bioactive compounds, are beyond a basic means to satisfy hunger or prevent food-deficiency diseases (Hasler 2000; Clydesdale 1999), which can bring optimal health and wellness to us.

Recently, researchers have become increasingly concerned with the chemopreventive activity of natural products from dietary and herbal plants for NAFLD prevention or treatment. Various bioactive compounds derived from fruits, vegetables, and edible plants are supposed to have health-promoting effects via anti-inflammatory, anti-oxidative, insulin sensitizer and weight-reducing properties. Several types of research have also been conducted in this regard. The therapeutic properties of these compounds are attributed to their effects on lowering blood glucose levels, improving lipid profile, hepatoprotective effect, and most importantly, without significant side effects (Pan et al. 2014; Sun Kim et al. 2012).

2.9. Cirrhosis

Liver cirrhosis is a final consequence of chronic liver damage characterized by progressive fibrosis, portal hypertension, and the development of regenerative nodules (Nusrat et al. 2014). Hepatic stellate cells (HSCs) activation is a fundamental phase in liver fibrogenesis, which induces the release of extracellular matrix components followed by differentiation of the HSC to a myofibroblast-like cell. Although the responsible agents for HSC activation are still not recognized, oxidative stress and various cytokines such as transforming growth factor β-1 and platelet-derived growth factor are involved. These procedures lead to overproduction of extracellular matrix (ECM) (Freidman 1993; George 2003; Bataller and Brenner 2005; Halliwell 2007; Aoyama et al. 2012).

According to World Health Organization (WHO) report, chronic diseases account for about half of worldwide death and diseases burden (Murray and Lopez 1996). Annually, nearly one million people in the world die from cirrhosis. The prevalence of liver cirrhosis has been estimated at 1%, and even 4.5-9.5% in the autopsy series (Everhart 2008; Lim and Kim 2008). Excessive alcohol consumption in western countries and hepatitis B virus in Iran are identified as the main reasons for liver cirrhosis (Saberifiroozi 2017). Cirrhosis causes numerous complications, and patients with cirrhosis have a decreased life expectancy (Garcia–Tsao 2001). In 2010, cirrhosis was introduced as the eighth leading cause of death in the United States (Murray et al. 2013). The cirrhosis prognosis depends on the severity of allied complications, including gastroesophageal varices, peritonitis and ascites, hepatic encephalopathy (HE), hepatorenal syndrome and other metabolic disorders (Fukui et al. 2016). Hepatic encephalopathy is one of the most severe complications in advanced cirrhosis in 30-45% of patients (Amodio et al. 2001; Saxena et al. 2002). Recent studies have revealed that treatment of underlying causes, especially in earlier stages of cirrhosis can improve or even reverse fibrosis, which rejects the traditional belief that cirrhosis is irreversible (Perez-Tamayo 1979; Benyon and Iredale 2000).

Conclusion

The gastrointestinal system is the main site of nutrient digestion and absorption and, therefore, has substantial effects on providing energy and body growth. It is apparent that gastrointestinal disorders, with increasing lifetime prevalence, have profound impacts on human health and reduce the quality of life. The human intestinal tract harbors a complex microbial community with a substantial role in human health. Alteration in the composition and function of the gut microbiota is associated with different gastrointestinal disorders, and current therapeutic strategies focus on maintaining a healthy gut community.

References

Ahmed, M. H., Barakat, S., and Almobarak, A. O. 2012. "Nonalcoholic fatty liver disease and cardiovascular disease: has the time come for cardiologists to be hepatologists?" *Journal of Obesity* 2012.

Amodio, P., Del Piccolo, F., Pettenò, E., Mapelli, D., Angeli, P., Iemmolo, R., Muraca, M., Musto, C., Gerunda, G., and Rizzo, C. 2001. "Prevalence and prognostic value of quantified electroencephalogram (EEG) alterations in cirrhotic patients." *Journal of Hepatology* 35 (1): 37-45.

Angulo, P. 2002. "Nonalcoholic fatty liver disease." *New England Journal of Medicine* 346 (16): 1221-1231.

Aoyama, T., Paik, Y. H., Watanabe, S., Laleu, B., Gaggini, F., Fioraso-Cartier, L., Molango, S., Heitz, F., Merlot, C., and Szyndralewiez, C. 2012. "Nicotinamide adenine dinucleotide phosphate oxidase in experimental liver fibrosis: GKT137831 as a novel potential therapeutic agent." *Hepatology* 56 (6): 2316-2327.

Argo, C. K., Northup, P. G., Al-Osaimi, A. M., and Caldwell, S. H. 2009. "Systematic review of risk factors for fibrosis progression in non-alcoholic steatohepatitis." *Journal of Hepatology* 51 (2): 371-379.

Aro, P., Storskrubb, T., Ronkainen, J., Bolling-Sternevald, E., Engstrand, L., Vieth, M., Stolte, M., Talley, N. J., and Agréus, L. 2006. "Peptic ulcer disease in a general adult population: the Kalixanda study: a random population-based study." *American Journal of Epidemiology* 163 (11): 1025-1034.

Assimakopoulos, S. F., Papageorgiou, I., and Charonis, A. 2011. "Enterocytes' tight junctions: from molecules to diseases." *World Journal of Gastrointestinal Pathophysiology* 2 (6): 123.

Atia, A. N., and Buchman, A. L. 2009. "Oral rehydration solutions in non-cholera diarrhea: a review." *The American Journal of Gastroenterology* 104 (10): 2596-2604.

Ballance, S., Sahlstrøm, S., Lea, P., Nagy, N. E., Andersen, P. V., Dessev, T., Hull, S., Vardakou, M., and Faulks, R. 2013. "Evaluation of gastric processing and duodenal digestion of starch in six cereal meals on the associated glycaemic response using an adult fasted dynamic gastric model." *European Journal of Nutrition* 52 (2): 799-812.

Bandyopadhyay, D., Biswas, K., Reiter, R., and Banerjee, R. 2001. "Gastric toxicity and mucosal ulceration induced by oxygen-derived reactive species: protection by melatonin." *Current Molecular Medicine* 1 (4): 501-513.

Barbut, F., Richard, A., Hamadi, K., Chomette, V., Burghoffer, B., and Petit, J.-C. 2000. "Epidemiology of recurrences or reinfections ofClostridium difficile-associated diarrhea." *Journal of Clinical Microbiology* 38 (6): 2386-2388.

Barrett, K. E., Malley, J., and Naglieri, C. 2006. *Gastrointestinal Physiology*. Lange Medical Books/McGraw-Hill.

Basu, S., Chatterjee, M., Ganguly, S., and Chandra, P. K. 2007. "Effect of Lactobacillus rhamnosus GG in persistent diarrhea in Indian children: a randomized controlled trial." *Journal of Clinical Gastroenterology* 41 (8): 756-760.

Bataller, R., and Brenner, D. A. 2005. "Liver fibrosis." *The Journal of Clinical Investigation* 115 (2): 209-218.

Battistella, M., and Alibhai, S. 2004. "Evaluation and treatment of constipation." *Geriatrics & Aging* 5 (10): 23-27.

Baumgart, D. C., and Carding, S. R. 2007. "Inflammatory bowel disease: cause and immunobiology." *The Lancet* 369 (9573): 1627-1640.

Bellentani, S., Scaglioni, F., Marino, M., and Bedogni, G. 2010. "Epidemiology of non-alcoholic fatty liver disease." *Digestive Diseases* 28 (1): 155-161.

Benyon, R., and Iredale, J. 2000. "Is liver fibrosis reversible?" *Gut* 46 (4): 443-446.

Bernstein, C. N., Loftus, E. V., Ng, S. C., Lakatos, P. L., and Moum, B. 2012. "Hospitalisations and surgery in Crohn's disease." *Gut* 61 (4): 622-629.

Biasi, F., Astegiano, M., Maina, M., Leonarduzzi, G., and Poli, G. 2011. "Polyphenol supplementation as a complementary medicinal approach to treating inflammatory bowel disease." *Current Medicinal Chemistry* 18 (31): 4851-4865.

Booijink, C. C., El-Aidy, S., Rajilić-Stojanović, M., Heilig, H. G., Troost, F. J., Smidt, H., Kleerebezem, M., De Vos, W. M., and Zoetendal, E. G. 2010. "High temporal and inter-individual variation detected in the human ileal microbiota." *Environmental Microbiology* 12 (12): 3213-3227.

Brandon, S. 2004. "Nutritional supplement having a unique soluble and insoluble fiber blend and method of making the same." *Google Patents.*

Brown, A. C., Rampertab, S. D., and Mullin, G. E. 2011. "Existing dietary guidelines for Crohn's disease and ulcerative colitis." *Expert Review of Gastroenterology & Hepatology* 5 (3): 411-425.

Browning, J. D., Szczepaniak, L. S., Dobbins, R., Horton, J. D., Cohen, J. C., Grundy, S. M., and Hobbs, H. H. 2004. "Prevalence of hepatic steatosis in an urban population in the United States: impact of ethnicity." *Hepatology* 40 (6): 1387-1395.

Caballería, L., Pera, G., Rodríguez, L., Auladell, M. A., Bernad, J., Canut, S., and Toran, P. 2012. "Metabolic syndrome and nonalcoholic fatty liver disease in a Spanish population: influence of the diagnostic criteria used." *European Journal of Gastroenterology & hepatology* 24 (9): 1007-1011.

Calam, J., and Baron, J. "Pathophysiology of duodenal and gastric ulcer and gastric cancer." *ABC OF THE UPPER GASTROINTESTINAL TRACT*: 19.

Camilleri, M., and Murray, J. A. 2008. "Diarrhea and constipation." *Harrisons Principles of Internal Medicine. Fauci, AS ed., McGraw Hill Medical, New York*: 245-255.

Casburn-Jones, A., and Farthing, M. 2004. "Management of infectious diarrhoea." *Gut* 53 (2): 296-305.

Catassi, C., Fabiani, E., Iacono, G., D'agate, C., Francavilla, R., Biagi, F., Volta, U., Accomando, S., Picarelli, A., and De Vitis, I. 2007. "A prospective, double-blind, placebo-controlled trial to establish a safe gluten threshold for patients with celiac disease." *The American journal of Clinical Nutrition* 85 (1): 160-166.

Chalasani, N., Younossi, Z., Lavine, J. E., Diehl, A. M., Brunt, E. M., Cusi, K., Charlton, M., and Sanyal, A. J. 2012. "The diagnosis and management of non-alcoholic fatty liver disease: practice guideline by the American Gastroenterological Association, American Association for the Study of Liver Diseases, and American College of Gastroenterology." *Gastroente-rology* 142 (7): 1592-1609.

Charytoniuk, T., Drygalski, K., Konstantynowicz-Nowicka, K., Berk, K., and Chabowski, A. 2017. "Alternative treatment methods attenuate the development of NAFLD: A review of resveratrol molecular mechanisms and clinical trials." *Nutrition* 34: 108-117.

Chu, A. 2014. "Antagonism by bioactive polyphenols against inflammation: a systematic view." *Inflammation & Allergy-Drug* 13 (1): 34-64.

Ciarlet, M., and Estes, M. K. 2001. "Interactions between rotavirus and gastrointestinal cells." *Current Opinion in Microbiology* 4 (4): 435-441.

Clydesdale, F. 1999. "ILSI North America food component reports." *Critical Reviews in Food Science and Nutrition* 39 (3): 203-316.

da Silva, M. S., Sánchez-Fidalgo, S., Talero, E., Cárdeno, A., da Silva, M. A., Villegas, W., Brito, A. R. M. S., and de La Lastra, C. A. 2010. "Anti-inflammatory intestinal activity of Abarema cochliacarpos (Gomes) Barneby & Grimes in TNBS colitis model." *Journal of Ethnopharmacology* 128 (2): 467-475.

de Lira Mota, K. S., Dias, G. E. N., Pinto, M. E. F., Luiz-Ferreira, Â., Monteiro Souza-Brito, A. R., Hiruma-Lima, C. A., Barbosa-Filho, J. M., and Batista, L. M. 2009. "Flavonoids with gastroprotective activity." *Molecules* 14 (3): 979-1012.

Dimidi, E., Christodoulides, S., Scott, S. M., and Whelan, K. 2017. "Mechanisms of Action of Probiotics and the Gastrointestinal Microbiota on Gut Motility and Constipation." *Advances in Nutrition* 8 (3): 484-494.

Drewes, A. M., Frφkjær, J. B., Larsen, E., Reddy, H., Arendt-Nielsen, L., and Gregersen, H. 2006. "Pain and mechanical properties of the rectum in patients with active ulcerative colitis." *Inflammatory Bowel Diseases* 12 (4): 294-303.

Drossman, D. A., Li, Z., Andruzzi, E., Temple, R. D., Talley, N. J., Thompson, W. G., Whitehead, W. E., Janssens, J., Funch-Jensen, P., and Corazziari, E. 1993. "US householder survey of functional gastrointestinal disorders." *Digestive Diseases and Sciences* 38 (9): 1569-1580.

Dubé, C., Rostom, A., Sy, R., Cranney, A., Saloojee, N., Garritty, C., Sampson, M., Zhang, L., Yazdi, F., and Mamaladze, V. 2005. "The prevalence of celiac disease in average-risk and at-risk Western European populations: a systematic review." *Gastroenterology* 128 (4): S57-S67.

Dyson, J. K., Anstee, Q. M., and McPherson, S. 2013. "Non-alcoholic fatty liver disease: a practical approach to diagnosis and staging." *Frontline Gastroenterology* 5 (3): 211-218.

Ekstedt, M., Franzén, L. E., Mathiesen, U. L., Thorelius, L., Holmqvist, M., Bodemar, G., and Kechagias, S. 2006. "Long-term follow-up of patients with NAFLD and elevated liver enzymes." *Hepatology* 44 (4): 865-873.

El-Serag, H. B., Sweet, S., Winchester, C. C., and Dent, J. 2013. "Update on the epidemiology of gastro-oesophageal reflux disease: a systematic review." *Gut* 63 (6); 871-880.

Everhart, J. 2008. "The burden of digestive diseases in the United States. US Department of Health and Human Services." *Public Health Service, National Institutes of Health, National Institute of Diabetes and Digestive and Kidney Diseases. Washington, DC: US Government Printing Office*: 111-114.

Everhart, J. E., Bryd-Holt, D., and Sonnenberg, A. 1998. "Incidence and risk factors for self-reported peptic ulcer diseasse in the United States." *American Journal of Epidemiology* 147 (6): 529-536.

Excellence, N. I. f. C. 2008. "Irritable bowel syndrome in adults: Diagnosis and management of irritable bowel syndrome in primary care." *London: National Institute for Clinical Excellence*.

Farre, R., and Tack, J. 2013. "Food and symptom generation in functional gastrointestinal disorders: physiological aspects." *The American Journal of Gastroenterology* 108 (5): 698-706.

Farzaei, M. H., Abdollahi, M., and Rahimi, R. 2015. "Role of dietary polyphenols in the management of peptic ulcer." *World Journal of Gastroenterol* 21 (21): 6499-517.

Fasano, A., and Catassi, C. 2001. "Current approaches to diagnosis and treatment of celiac disease: an evolving spectrum." *Gastroenterology* 120 (3): 636-651.

Fernández, A., Barreiro-de Acosta, M., Vallejo, N., Iglesias, M., Carmona, A., González-Portela, C., Lorenzo, A., and Domínguez-Muñoz, J. E. 2012. "Complementary and alternative medicine in inflammatory bowel disease patients: frequency and risk factors." *Digestive and Liver Disease* 44 (11): 904-908.

Freidman, S. 1993. "The cellular basis of hepatic fibrosis." The *New England Journal of Medicine* 328: 1828-1835.

Fukui, H., Saito, H., Ueno, Y., Uto, H., Obara, K., Sakaida, I., Shibuya, A., Seike, M., Nagoshi, S., and Segawa, M. 2016. "Evidence-based clinical practice guidelines for liver cirrhosis 2015." *Journal of Gastroenterology* 51 (7): 629-650.

Garcia–Tsao, G. 2001. "Current management of the complications of cirrhosis and portal hypertension: variceal hemorrhage, ascites, and spontaneous bacterial peritonitis." *Gastroenterology* 120 (3): 726-748.

George, J. 2003. "Ascorbic acid concentrations in dimethylnitrosamine-induced hepatic fibrosis in rats." *Clinica Chimica Acta* 335 (1-2): 39-47.

Goel, R., and Sairam, K. 2002. "Anti-ulcer drugs from indigenous sources with emphasis on Musa sapientum, tamrahbasma, Asparagus racemosus and Zingiber officinale." *Indian Journal of Pharmacology* 34 (2): 100-110.

Golding, M., and Wooster, T. J. 2010. "The influence of emulsion structure and stability on lipid digestion." *Current Opinion in Colloid & Interface Science* 15 (1-2): 90-101.

Green, P. H., Lebwohl, B., and Greywoode, R. 2015. "Celiac disease." *Journal of Allergy and Clinical Immunology* 135 (5): 1099-1106.

Gregorio, G. V., Dans, L. F., Cordero, C. P., and Panelo, C. A. 2007. "Zinc supplementation reduced cost and duration of acute diarrhea in children." *Journal of Clinical Epidemiology* 60 (6): 560-566.

Gupta, S. C., Tyagi, A. K., Deshmukh-Taskar, P., Hinojosa, M., Prasad, S., and Aggarwal, B. B. 2014. "Downregulation of tumor necrosis factor and other proinflammatory biomarkers by polyphenols." *Archives of Biochemistry and Biophysics* 559: 91-99.

Hajiaghamohammadi, A. A., Ziaee, A., and Samimi, R. 2012. "The Efficacy of Licorice Root Extract in Decreasing Transaminase Activities in Non-alcoholic Fatty Liver Disease: A Randomized Controlled Clinical Trial." *Phytotherapy Research* 26 (9): 1381-1384.

Hall, E. 2009. "Canine diarrhoea: a rational approach to diagnostic and therapeutic dilemmas." *In Practice* 31 (1): 8-16.

Halliwell, B. 2007. "Oxidative stress and cancer: have we moved forward?" *Biochemical Journal* 401 (1): 1-11.

Hanif Palla, A., and Gilani, A. H. 2015. "Dual effectiveness of Flaxseed in constipation and diarrhea: Possible mechanism." *J Ethnopharmacol* 169: 60-8.

Hasler, C. M. 1998. "Functional foods: their role in disease prevention and health promotion." *Food Technology-Champaign Then Chicago* 52: 63-147.

Hasler, C. M. 2000. "The changing face of functional foods." *Journal of the American College of Nutrition* 19 (5 Suppl): 499S-506S.

Hasler, C. M., and Brown, A. C. 2009. "Position of the American Dietetic Association: functional foods." *Journal of the American Dietetic Association* 109 (4): 735-46.

He, J., Magnuson, B. A., and Giusti, M. M. 2005. "Analysis of anthocyanins in rat intestinal contents impact of anthocyanin chemical structure on fecal excretion." *Journal of Agricultural and Food Chemistry* 53 (8): 2859-2866.

Hoekstra, J., Szajewska, H., Zikri, M. A., Micetic-Turk, D., Weizman, Z., Papadopoulou, A., Guarino, A., Dias, J., and Oostvogels, B. 2004. "Oral rehydration solution containing a mixture of non-digestible carbohydrates in the treatment of acute diarrhea: a multicenter randomized placebo controlled study on behalf of the ESPGHAN working group on intestinal infections." *Journal of Pediatric Gastroenterology and Nutrition* 39 (3): 239-245.

Hold, G. L., Smith, M., Grange, C., Watt, E. R., El-Omar, E. M., and Mukhopadhya, I. 2014. "Role of the gut microbiota in inflammatory bowel disease pathogenesis: what have we learnt in the past 10 years?" *World Journal of Gastroenterology: WJG* 20 (5): 1192.

Horwitz, B. J., and Fisher, R. S. 2001. "The irritable bowel syndrome." *New England Journal of Medicine* 344 (24): 1846-1850.

Hou, J. K., Abraham, B., and El-Serag, H. 2011. "Dietary intake and risk of developing inflammatory bowel disease: a systematic review of the literature." *The American Journal of Gastroenterology* 106 (4): 563.

Isolauri, E., Kaila, M., Mykkänen, H., Ling, W. H., and Salminen, S. 1994. "Oral bacteriotherapy for viral gastroenteritis." *Digestive Diseases and Sciences* 39 (12): 2595-2600.

Iwańczak, F., and Iwańczak, B. 2012. "New guidelines for diagnosis and treatment of coeliac disease in children and adolescents." *Gastroenterology Review/Przegląd Gastroenterologiczny* 7 (4): 185-191.

Johnson, L., Ghishan, F., Kaunitz, J., Merchant, J., Said, H., and Wood, J. 2012. *Physiology of the Gastrointestinal Tract*. Elsevier Inc.

Jones, M. P. 2006. "The role of psychosocial factors in peptic ulcer disease: beyond Helicobacter pylori and NSAIDs." *Journal of Psychosomatic Research* 60 (4): 407-412.

Kalichman, L., Livshits, G., and Kobyliansky, E. 2006. "Indices of body composition and chronic morbidity: A cross-sectional study of a rural population in central Russia." *American Journal of Human Biology* 18 (3): 350-358.

Kani, A. H., Alavian, S. M., Esmaillzadeh, A., Adibi, P., Haghighatdoost, F., and Azadbakht, L. 2017. "Effects of a Low-Calorie, Low-Carbohydrate Soy Containing Diet on Systemic Inflammation Among Patients with Nonalcoholic Fatty Liver Disease: A Parallel Randomized Clinical Trial." *Hormon and Metaboilc Research* 49 (9): 687-692.

Kassinen, A., Krogius-Kurikka, L., Mäkivuokko, H., Rinttilä, T., Paulin, L., Corander, J., Malinen, E., Apajalahti, J., and Palva, A. 2007. "The fecal microbiota of irritable bowel syndrome patients differs significantly from that of healthy subjects." *Gastroenterology* 133 (1): 24-33.

Kaulmann, A., and Bohn, T. 2016. "Bioactivity of polyphenols: Preventive and adjuvant strategies toward reducing inflammatory bowel diseases—promises, perspectives, and pitfalls." *Oxidative Medicine and Cellular Longevity* 2016.

Kellow, J. E., Azpiroz, F., Delvaux, M., Gebhart, G., Mertz, H. R., Quigley, E. M., and Smout, A. J. 2006. "Applied principles of neurogastroenterology: physiology/motility sensation." *Gastroenterology* 130 (5): 1412-1420.

Kim, E. R., and Rhee, P.-L. 2012. "How to interpret a functional or motility test-colon transit study." *Journal of Neurogastroenterology and Motility* 18 (1): 94.

King, T., Elia, M., and Hunter, J. 1998. "Abnormal colonic fermentation in irritable bowel syndrome." *The Lancet* 352 (9135): 1187-1189.

Klauser, A., Schindlbeck, N., and Müller-Lissner, S. 1990. "Symptoms in gastro-oesophageal reflux disease." *The Lancet* 335 (8683): 205-208.

Kosek, M., Bern, C., and Guerrant, R. L. 2003. "The global burden of diarrhoeal disease, as estimated from studies published between 1992 and 2000." *Bulletin of the World Health Organization* 81 (3): 197-204.

Kumar, R., Prakash, S., Chhabra, S., Singla, V., Madan, K., Gupta, S. D., Panda, S. K., Khanal, S., and Acharya, S. K. 2012. "Association of pro-inflammatory cytokines, adipokines & oxidative stress with insulin resistance & non-alcoholic fatty liver disease." *The Indian Journal of Medical Research* 136 (2): 229.

Kurata, J. H., and Nogawa, A. N. 1997. "Meta-analysis of risk factors for peptic ulcer: nonsteroidal antiinflammatory drugs, Helicobacter pylori, and smoking." *Journal of Clinical Gastroenterology* 24 (1): 2-17.

Kyne, L., and Kelly, C. 2001. "Recurrent Clostridium difficilediarrhoea." *Gut* 49 (1): 152-153.

LaBrecque, D. R., Abbas, Z., Anania, F., Ferenci, P., Khan, A. G., Goh, K. L., Hamid, S. S., Isakov, V., Lizarzabal, M., and Peñaranda, M. M. 2014. "World Gastroenterology Organisation global guidelines: Nonalcoholic fatty liver disease and nonalcoholic steatohepatitis." *Journal of Clinical Gastroenterology* 48 (6): 467-473.

Lakatos, P. L., Czegledi, Z., David, G., Kispal, Z., Kiss, L. S., Palatka, K., Kristof, T., Nagy, F., Salamon, A., and Demeter, P. 2010. "Association of adherence to therapy and complementary and alternative medicine use with demographic factors and disease phenotype in patients with inflammatory bowel disease." *Journal of Crohn's and Colitis* 4 (3): 283-290.

Larsen, S., Bendtzen, K., and Nielsen, O. H. 2010. "Extraintestinal manifestations of inflammatory bowel disease: epidemiology, diagnosis, and management." *Annals of Medicine* 42 (2): 97-114.

Lazo, M., and Clark, J. M. 2008. "The epidemiology of nonalcoholic fatty liver disease: a global perspective." Seminars in liver disease.

Lazo, M., Hernaez, R., Eberhardt, M. S., Bonekamp, S., Kamel, I., Guallar, E., Koteish, A., Brancati, F. L., and Clark, J. M. 2013. "Prevalence of nonalcoholic fatty liver disease in the United States: the Third National Health and Nutrition Examination Survey, 1988–1994." *American Journal of Epidemiology* 178 (1): 38-45.

Lee, A., and Newman, J. M. 2003. "Celiac diet: its impact on quality of life." *Journal of the American Dietetic Association* 103 (11): 1533-1535.

Lee, D., Albenberg, L., Compher, C., Baldassano, R., Piccoli, D., Lewis, J. D., and Wu, G. D. 2015. "Diet in the Pathogenesis and Treatment of Inflammatory BoweláDiseases." *Gastroenterology* 148 (6): 1087-1106.

Lichtenstein, G. R., Rutgeerts, P., Sandborn, W. J., Sands, B. E., Diamond, R. H., Blank, M., Montello, J., Tang, L., Cornillie, F., and Colombel, J.-F. 2012. "A pooled analysis of infections, malignancy, and mortality in infliximab-and immunomodulator-treated adult patients with inflammatory bowel disease." *The American Journal of Gastroentero-logy* 107 (7): 1051.

Lim, Y.-S., and Kim, W. R. 2008. "The global impact of hepatic fibrosis and end-stage liver disease." *Clinics in Liver Disease* 12 (4): 733-746.

Liu, L., Johnson, H. L., Cousens, S., Perin, J., Scott, S., Lawn, J. E., Rudan, I., Campbell, H., Cibulskis, R., and Li, M. 2012. "Global, regional, and national causes of child mortality: an updated systematic analysis for 2010 with time trends since 2000." *The Lancet* 379 (9832): 2151-2161.

Locke, G. r., Talley, N. J., Fett, S. L., Zinsmeister, A. R., and Melton, L. r. 1997. "Prevalence and clinical spectrum of gastroesophageal reflux: a population-based study in Olmsted County, Minnesota." *Gastroentero-logy* 112 (5): 1448-1456.

Ma, X., and Li, Z. 2006. "Pathogenesis of nonalcoholic steatohepatitis (NASH)." *Journal of Digestive Diseases* 7 (1): 7-11.

Martins, L. C., de Oliveira Corvelo, T. C., Oti, H. T., and dos Santos Barile, K. A. 2002. "Soroprevalência de anticorpos contra o antígeno CagA do Helicobacter pyloriem pacientes com úlcera gástrica na região Norte do Brasil." *Revista da Sociedade Brasileira de Medicina Tropical* 35 (4): 307-310.

Mason, H. J., Serrano-Ikkos, E., and Kamm, M. A. 2000. "Psychological morbidity in women with idiopathic constipation." *The American Journal of Gastroenterology* 95 (10): 2852-2857.

Mason, J., and Hungin, A. 2005. "Gastro-oesophageal reflux disease–the health economic implications." *Alimentary Pharmacology & Therapeutics* 22 (s1): 20-31.

Mason, K. L., Huffnagle, G. B., Noverr, M. C., and Kao, J. Y. 2008. "Overview of gut immunology." In *GI Microbiota and Regulation of The Immune System*, 1-14. Springer.

Maximos, M., Bril, F., Portillo Sanchez, P., Lomonaco, R., Orsak, B., Biernacki, D., Suman, A., Weber, M., and Cusi, K. 2015. "The role of liver fat and insulin resistance as determinants of plasma aminotransferase elevation in nonalcoholic fatty liver disease." *Hepatology* 61 (1): 153-160.

Mayer, E. A. 2011. "Gut feelings: the emerging biology of gut–brain communication." *Nature Reviews Neuroscience* 12 (8): 453.

Michelotti, G. A., Machado, M. V., and Diehl, A. M. 2013. "NAFLD, NASH and liver cancer." *Nature Reviews Gastroenterology and Hepatology* 10 (11): 656-665.

Mrukowicz, J., Krobicka, B., Duplaga, M., Kowalska-Duplaga, K., Domanski, J., Kantecki, M., Iwañczak, F., and Pytrus, T. 1999. "Epidemiology and impact of rotavirus diarrhoea in Poland." *Acta Paediatrica* 88 (s426): 53-60.

Munos, M. K., Walker, C. L. F., and Black, R. E. 2010. "The effect of oral rehydration solution and recommended home fluids on diarrhoea mortality." *International Journal of Epidemiology* 39 (suppl_1): i75-i87.

Murray, C. J., Abraham, J., Ali, M. K., Alvarado, M., Atkinson, C., Baddour, L. M., Bartels, D. H., Benjamin, E. J., Bhalla, K., and Birbeck, G. 2013. "The state of US health, 1990-2010: burden of diseases, injuries, and risk factors." *Jama* 310 (6): 591-606.

Murray, C. J., and Lopez, A. D. 1996. "Evidence-based health policy--lessons from the Global Burden of Disease Study." *Science* 274 (5288): 740-743.

Musso, G., Gambino, R., and Cassader, M. 2009. "Recent insights into hepatic lipid metabolism in non-alcoholic fatty liver disease (NAFLD)." *Progress in Lipid Research* 48 (1): 1-26.

Nakamuta, M., Kohjima, M., Higuchi, N., Kato, M., Kotoh, K., Yoshimoto, T., Yada, M., Yada, R., Takemoto, R., and Fukuizumi, K. 2008. "The significance of differences in fatty acid metabolism between obese and non-obese patients with non-alcoholic fatty liver disease." *International Journal of Molecular Medicine* 22 (5): 663-667.

Narayanappa, D. 2008. "Randomized double blinded controlled trial to evaluate the efficacy and safety of Bifilac in patients with acute viral diarrhea." *Indian Journal of Pediatrics* 75 (7): 709-713.

National Institute of Diabetes and Digestive and Kidney Diseases (NDDIK). 2010. "Irritable Bowel Syndrome." http://digestive.niddk.nih.gov/ ddiseases/pubs/ibs/index. htm.

Neuschwander-Tetri, B. A., Clark, J. M., Bass, N. M., Van Natta, M. L., Unalp-Arida, A., Tonascia, J., Zein, C. O., Brunt, E. M., Kleiner, D. E., and McCullough, A. J. 2010. "Clinical, laboratory and histological associations in adults with nonalcoholic fatty liver disease." *Hepatology* 52 (3): 913-924.

Nobaek, S., Johansson, M.-L., Molin, G., Ahrné, S., and Jeppsson, B. 2000. "Alteration of intestinal microflora is associated with reduction in abdominal bloating and pain in patients with irritable bowel syndrome." *The American Journal of Gastroenterology* 95 (5): 1231-1238.

Nusrat, S., Khan, M. S., Fazili, J., and Madhoun, M. F. 2014. "Cirrhosis and its complications: evidence-based treatment." *World Journal of Gastroenterology: WJG* 20 (18): 5442.

Ong, J. P., and Younossi, Z. M. 2007. "Epidemiology and natural history of NAFLD and NASH." *Clinics in Liver Disease* 11 (1): 1-16.

Pan, M. H., Lai, C. S., Tsai, M. L., and Ho, C. T. 2014. "Chemoprevention of nonalcoholic fatty liver disease by dietary natural compounds." *Molecular Nutrition and Food Research* 58 (1): 147-71.

Pare, P., Ferrazzi, S., Thompson, W., Irvine, E., and Rance, L. 2001. "An epidemiological survey of constipation in Canada: definitions, rates, demographics, and predictors of health care seeking." *The American Journal of Gastroenterology* 96 (11): 3130-3137.

Parkes, G. C., Brostoff, J., Whelan, K., and Sanderson, J. D. 2008. "Gastrointestinal microbiota in irritable bowel syndrome: their role in its pathogenesis and treatment." *The American Journal of Gastroenterology* 103 (6): 1557.

Passariello, A., Terrin, G., De Marco, G., Cecere, G., Ruotolo, S., Marino, A., Cosenza, L., Tardi, M., Nocerino, R., and Canani, R. B. 2011. "Efficacy of a new hypotonic oral rehydration solution containing zinc and prebiotics in the treatment of childhood acute diarrhea: a randomized controlled trial." *The Journal of Pediatrics* 158 (2): 288-292. e1.

Peppas, G., Alexiou, V. G., Mourtzoukou, E., and Falagas, M. E. 2008. "Epidemiology of constipation in Europe and Oceania: a systematic review." *BMC Gastroenterology* 8 (1): 5.

Perez-Tamayo, R. 1979. "Cirrhosis of the liver: a reversible disease?" *Pathology Annual* 14: 183-213.

Podolsky, D. K. 2002. "Inflammatory bowel disease." The *New England Journal of Medicine* 347 (6): 417-29.

Putignani, L., Del Chierico, F., Vernocchi, P., Cicala, M., Cucchiara, S., Dallapiccola, B., and Group, D. S. 2015. "Gut Microbiota Dysbiosis as Risk and Premorbid Factors of IBD and IBS Along the Childhood–Adulthood Transition." *Inflammatory Bowel Diseases* 22 (2): 487-504.

Reis, N. T. 2003. *Nutrição Clínica Sistema Digestório*. [*Clinical Nutrition Digestive System*]. Rubio.

Riddle, M. S., Murray, J. A., and Porter, C. K. 2012. "The incidence and risk of celiac disease in a healthy US adult population." *The American Journal of Gastroenterology* 107 (8): 1248-1255.

Rocchi, A., Benchimol, E. I., Bernstein, C. N., Bitton, A., Feagan, B., Panaccione, R., Glasgow, K. W., Fernandes, A., and Ghosh, S. 2012. "Inflammatory bowel disease: a Canadian burden of illness review." *Canadian Journal of Gastroenterology and Hepatology* 26 (11): 811-817.

Romero, Y., Cameron, A. J., Locke, G. r., Schaid, D. J., Slezak, J. M., Branch, C. D., and Melton, L. 1997. "Familial aggregation of gastroesophageal reflux in patients with Barrett's esophagus and esophageal adenocarcinoma." *Gastroenterology* 113 (5): 1449-1456.

Romier, B., Schneider, Y.-J., Larondelle, Y., and During, A. 2009. "Dietary polyphenols can modulate the intestinal inflammatory response." *Nutrition Reviews* 67 (7): 363-378.

Ronkainen, J., Aro, P., Storskrubb, T., Lind, T., Bolling-Sternevald, E., Junghard, O., Talley, N., and Agreus, L. 2006. "Gastro-oesophageal reflux symptoms and health-related quality of life in the adult general population–the Kalixanda study." *Alimentary Pharmacology & Therapeutics* 23 (12): 1725-1733.

Rubio-Tapia, A., Hill, I. D., Kelly, C. P., Calderwood, A. H., and Murray, J. A. 2013. "ACG clinical guidelines: diagnosis and management of celiac disease." *The American Journal of Gastroenterology* 108 (5): 656-676.

Saberifiroozi, M. 2017. "Improving Quality of Care in Patients with Liver Cirrhosis." *Middle East Journal of Digestive Diseases* 9 (4): 189-200.

Sahebkar, A. 2011. "Citrus auraptene: A potential multifunctional therapeutic agent for nonalcoholic fatty liver disease." *Annals of Hepatology* 10 (4): 575-7.

Sandler, R. S., and Drossman, D. A. 1987. "Bowel habits in young adults not seeking health care." *Digestive Diseases and Sciences* 32 (8): 841-845.

Saxena, N., Bhatia, M., Joshi, Y. K., Garg, P. K., Dwivedi, S. N., and Tandon, R. K. 2002. "Electrophysiological and neuropsychological tests for the diagnosis of subclinical hepatic encephalopathy and prediction of overt encephalopathy." *Liver International* 22 (3): 190-197.

Schubert, M. L. 2008. "Hormonal regulation of gastric acid secretion." *Current Gastroenterology Reports* 10 (6): 523-527.

Schuppan, D., Junker, Y., and Barisani, D. 2009. "Celiac disease: from pathogenesis to novel therapies." *Gastroenterology* 137 (6): 1912-1933.

See, J., and Murray, J. A. 2006. "Gluten-free diet: the medical and nutrition management of celiac disease." *Nutrition in Clinical Practice* 21 (1): 1-15.

Shih, D. Q., and Targan, S. R. 2008. "Immunopathogenesis of inflammatory bowel disease." *World Journal of Gastroenterology: WJG* 14 (3): 390.

Söderholm, J. D., Streutker, C., Yang, P., Paterson, C., Singh, P., McKay, D., Sherman, P., Croitoru, K., and Perdue, M. 2004. "Increased epithelial uptake of protein antigens in the ileum of Crohn's disease mediated by tumour necrosis factor α." *Gut* 53 (12): 1817-1824.

Sonnenberg, A., and Everhart, J. E. 1996. "The prevalence of self-reported peptic ulcer in the United States." *American Journal of Public Health* 86 (2): 200-205.

Stevens, L., and Rashid, M. 2008. "Gluten-free and regular foods: a cost comparison." *Canadian Journal of Dietetic Practice and Research* 69 (3): 147-150.

Sun Kim, M., Kung, S., Grewal, T., and D Roufogalis, B. 2012. "Methodologies for investigating natural medicines for the treatment of nonalcoholic fatty liver disease (NAFLD)." *Current Pharmaceutical Biotechnology* 13 (2): 278-291.

Sung, J. J., Tsoi, K. K., Ma, T. K., Yung, M.-Y., Lau, J. Y., and Chiu, P. W. 2010. "Causes of mortality in patients with peptic ulcer bleeding: a prospective cohort study of 10,428 cases." *The American Journal of Gastroenterology* 105 (1): 84-89.

Suter, M., Dorta, G., Giusti, V., and Calmes, J. 2004. "Gastro-esophageal reflux and esophageal motility disorders in morbidly obese patients." *Obesity Surgery* 14 (7): 959-966.

Taylor, D., and Echeverria, P. 1993. "Diarrhoeal disease: current concepts and future challenges. Molecular biological approaches to the epidemiology of diarrhoeal diseases in developing countries." *Transactions of the Royal Society of Tropical Medicine and Hygiene* 87: 3-5.

Tennyson, C. A., Simpson, S., Lebwohl, B., Lewis, S., and Green, P. H. 2013. "Interest in medical therapy for celiac disease." *Therapeutic Advances in Gastroenterology* 6 (5): 358-364.

Toneto, M., Oliveira, F., and Lopes, M. 2011. "Evolução histórica da úlcera péptica: da etiologia ao tratamento." *Scientia Medica* 21 (1): 23-30.

Toruner, M., Loftus, E. V., Harmsen, W. S., Zinsmeister, A. R., Orenstein, R., Sandborn, W. J., Colombel, J. F., and Egan, L. J. 2008. "Risk factors for opportunistic infections in patients with inflammatory bowel disease." *Gastroenterology* 134 (4): 929-936.

Trudgill, N. J., Kapur, K. C., and Riley, S. A. 1999. "Familial clustering of reflux symptoms." *The American Journal of Gastroenterology* 94 (5): 1172-1178.

Valle, D. 2005. "Peptic ulcer diseases and related disorders." *Harrison's Principles Of Internal Medicine* 16: 1746-1762.

van den Bogert, B., de Vos, W. M., Zoetendal, E. G., and Kleerebezem, M. 2011. "Microarray analysis and barcoded pyrosequencing provide consistent microbial profiles depending on the source of human intestinal samples." *Applied and Environmental Microbiology* 77 (6): 2071-2080.

Wahlqvist, P., Reilly, M., and Barkun, A. 2006. "Systematic review: the impact of gastro-oesophageal reflux disease on work productivity." *Alimentary Pharmacology & Therapeutics* 24 (2): 259-272.

Watson, R. R., and Preedy, V. R. 2012. *Bioactive Food as Dietary Interventions for Liver and Gastrointestinal Disease*. Academic Press.

Whitehead, W., Chaussade, S., Corazziari, E., and Kumar, D. 1991. "Report of an international workshop on management of constipation." *Gastroenterology International* 4 (3): 99-113.

Willemijntje, A., and Pankaj, J. 2001. "Agents used for control of gastric acidity and treatment of peptic ulcers and gastroesophageal reflux disease." *The Pharmacological Basis of Therapeutics. GH Joel & EL Lee, Eds*: 1010-1011.

Woda, A., Mishellany-Dutour, A., Batier, L., François, O., Meunier, J., Reynaud, B., Alric, M., and Peyron, M. 2010. "Development and validation of a mastication simulator." *Journal of Biomechanics* 43 (9): 1667-1673.

Wolf, J. 2012. *A Woman's Guide to a Healthy Stomach: Taking Control of Your Digestive Health*. Harlequin.

Yasutake, K., Nakamuta, M., Shima, Y., Ohyama, A., Masuda, K., Haruta, N., Fujino, T., Aoyagi, Y., Fukuizumi, K., and Yoshimoto, T. 2009. "Nutritional investigation of non-obese patients with non-alcoholic fatty liver disease: the significance of dietary cholesterol." *Scandinavian Journal of Gastroenterology* 44 (4): 471-477.

Chapter 2

The Role of Microbiota on the Health Status of the Gastrointestinal Tract

Abstract

The human gut harbors numerous microbial species that affect the host physiology, metabolisms, nutrition and immune system. The gut microbiota consists of bacteria, archaea, fungi, protozoa and viruses. Several factors such as genetics, environmental factors, diet and disease can alter the composition of gut microbiota known as "dysbiosis" which is associated with the incidence of various diseases such as irritable bowel syndrome, Crohn's disease, ulcerative colitis and colorectal cancer. In this respect, multiple studies have focused on efficient approaches to restore the healthy gut community. This chapter reviews the gut microbiota composition, the effect of diet on the composition, and the association between gut microbiota and different gastrointestinal diseases.

Keywords: diet, dysbiosis, gastrointestinal tract, gut microbiota

1. Introduction

The microbiota in the human gut plays an essential role in the balance between health and disease; there is a great symbiosis between the human host and its gut microbiota. It has been indicated that gut microbiota affects the immune system, physiology, nutrition and host metabolism (Guinane and Cotter 2013).

The gut microbiota consists of bacteria, archaea, fungi, protozoa and viruses. The number of inhabitant microorganisms in the gastrointestinal (GI) tract has been estimated to be more than 10^{14}; that includes ~10 times more bacterial cells than the number of human cells and over 100 times the amount of genomic content (microbiome) as the human genome (Thursby and Juge 2017). However, it has been inferred that the ratio is indeed closer to 1:1 (Sender, Fuchs, and Milo 2016). The predominant bacterial species in the human GI tract include three phyla: the phylum Bacteroidetes, the phylum

Firmicutes and the phylum Actinobacteria. Other bacteria including lactobacilli, streptococci and *Escherichia coli* are found in small numbers (Rajoka et al. 2017). In general, various microbial communities are colonized in different parts of the GI tract; Gram-positive bacteria are prevalent in the small intestine, while Gram-negative bacteria are predominant in the large intestine (Ghoshal and Ghoshal 2017). Several factors such as genetics, environmental factors, diet and disease can influence the composition of gut microbiota known as "dysbiosis" (Shukla et al. 2017). This status would lead to a series of diseases including obesity, diabetes, inflammatory bowel disease (IBD), ulcerative colitis and Crohn's disease (Guinane and Cotter 2013). It has been unraveled that food components can influence the composition of microbiota and production of bacteria-derived metabolites and, consequently, the host's health status.

2. Gut Microbiota Composition

Microbiota is developed after birth. The mode of delivery also influences microbiota composition. In vaginally born infants, *Lactobacillus*, *Prevotella*, and *Sneathia* spp. are dominant, while in infants delivered by Cesarean section, *Staphylococcus*, *Corynebacterium* and *Propionibacterium* are dominant (Dominguez-Bello et al. 2010). During the first year of life, the gut microbiota begins to modify toward a distinct profile similar to adults, but still unique for each infant (Palmer et al. 2007). By around 2.5 years of age, the infant microbiota's composition, diversity, and functional capabilities resemble those of adult microbiota (Rodríguez et al. 2015). In addition to delivery mode, host genetics, geographical location, lifestyle, diet and infections can affect the composition of the microbial community over the years.

The microbiota population is different in different parts of the GI tract, with the lowest concentration in the stomach and the highest number in the colon (Dieterich, Schink, and Zopf 2018). The stomach contains 10^1 bacteria per gram content with an increasing manner in the duodenum (10^3/g), jejunum (10^4/g), ileum (10^7/g), and colon (10^{12}/g) (Sekirov et al. 2010). Most microorganisms in the human gut are anaerobic, belonging to phyla Bacteroidetes, Firmicutes, and Proteobacteria. Other gut bacteria with a minor portion in the healthy gut frequently belong to phyla Actinobacteria, Verrumicrobia, Acidobacteria, or Fusobacteria (Dieterich et al. 2019).

In a healthy stomach, five main phyla, including Firmicutes, Bacteroidetes, Actinobacteria, Fusobacteria and Proteobacteria, and the bacterial genera *Prevotella*, *Streptococcus*, *Veillonella*, *Rothia* and *Haemophilus,* are dominant. In the small intestine, bile acids, pancreatic enzymes and antimicrobial agents in the duodenum prevent bacteria colonization and the predominant phyla are *Firmicutes* and Actinobacteria. In the jejunum, the microbial population comprises Gram-positive aerobes and facultative anaerobes including lactobacilli, enterococci and streptococci.

In the ileum, aerobic species are predominant while in the distal part of the ileum close to the ileocecal valve, anaerobes and Gram-negative organisms are found in great numbers (El Aidy, Van Den Bogert, and Kleerebezem 2015). The main bacterial phyla in the large intestine are *Firmicutes* and Bacteroidetes (Mariat et al. 2009). In the lumen, genera such as *Bacteroides*, *Bifdobacterium*, *Streptococcus*, *Enterobacteriaceae*, *Enterococcus*, *Clostridium*, *Lactobacillus* and *Ruminococcus* are predominant, whereas *Clostridium*, *Lactobacillus*, *Enterococcus* and *Akkermansia* are associated with the mucosa (Hollister, Gao, and Versalovic 2014).

Although the bacteria are the most common and dominant microorganisms in gut microbiota, the intestine also contains archaea, bacteriophages, viruses, unicellular eukaryotes, and fungi that contribute significantly to the health status of the microbial community (Norman et al. 2015). In this respect, bacteriophages are of great importance due to their effect on their bacterial hosts' survival, reproduction, composition, and functionality. It has been reported that a decrease in the bacteriophages population led to gastrointestinal complaints (Manrique et al. 2016). The other residents in the human gut are fungi, including the genera *Aspergillus*, *Candida*, *Cryptococcus*, and *Penicillium*, which consist of 0.2–0.3% of microbiota (Forbes, Van Domselaar, and Bernstein 2016). The relation of the intestinal mycobiota and diseases has not yet been elucidated, but changing the fungal patterns were associated with irritable bowel syndrome and inflammatory bowel diseases (Chiaro et al. 2017; Sokol et al. 2017).

3. Effect of Diet on Gut Microbiota

Different micro- and macronutrients in the diet can modulate gut microbiota composition and influence the host's health. Carbohydrates are the most common energy source for humans and significantly affect the composition of gut microbiota. The plant-derived carbohydrates known as dietary fibers that are not digested in the upper GI tract are fermented by bacterial species in the gut, resulting in various metabolites such as short-chain fatty acids (SCFAs), CO_2, and H_2 (Danneskiold-Samsøe et al. 2019). SCFAs play a pivotal role in the regulation of host metabolism. They can bind to the G-protein coupled receptors (GPCRs) involved in regulating processes, including glucose homeostasis and lipid metabolism (Canfora, Jocken, and Blaak 2015; Morrison and Preston 2016). It has been reported that SCFAs can modulate gene expression via histone deacetylase inhibition (Koh et al. 2016). Moreover, SCFAs are considered an important energy source for the intestinal epithelial cells and improve the gut barrier function (Liu et al. 2018). Evidence from clinical trials has revealed that a diet enriched with fibers increased the abundance of beneficial microbiota such as *Bifidobacterium*, *Lactobacillus*, *Akkermansia*, *Fecalibacterium*, *Roseburia*, *Bacteroides*, and *Prevotella* species (Yang et al. 2020). Furthermore, a decrease of Firmicutes to Bacteroidetes ratio and improvement of gut microbial diversity was observed by consuming a high-fiber diet (Holscher et al. 2015).

Polyphenols, including flavonoids, phenolic acids, stilbenes and lignans are suggested to modulate gut microbiota. These compounds prevent pathogenic microorganisms such as *Helicobacter pylori* and *Staphylococcus* spp. and increase the population of beneficial bacteria such as *Lactobacillus*, *Bifidobacterium*, *Akkermansia*, and *Fecalibacterium* spp. (Parkar, Stevenson, and Skinner 2008; Selma, Espin, and Tomas-Barberan 2009). These beneficial effects are mainly attributed to the prebiotic activity of polyphenols (Santino et al. 2017). In a study conducted by (Zhao et al. 2017), administration of quercetin and resveratrol (30 and 15 mg per kg body weight per day, respectively) to Wistar rats showed an effect on gut microbiota. The abundance of *Firmicutes* and the ratio of *Firmicutes* to *Bacteroidetes* decreased. Also, a decrease in the relative abundance of *Desulfovibrionaceae*, *Acidaminococcaceae*, *Coriobacteriaceae*, *Bilophila*, *Lachnospiraceae* and its genus *Lachnoclostridium*, which were reported to be potentially associated with diet-induced obesity, was observed. *In vitro* investigation of the effect of polyphenols from green tea, oolong tea and black tea on human intestinal microbiota revealed the proliferation of *Bacteroides-Prevotella* and *Clostridium histolyticum* groups were restricted while the *Bifidobacterium* and *Lactobacillus-Enterococcus* spp. and SCFAs production significantly increased (Sun et al. 2018).

It has been highlighted that the level of fat and its type have remarkable impacts on gut microbiota composition. Lowering the abundance of beneficial microbes such as *Bifidobacterium* and *Fecalibacterium* with a diet enriched with saturated fat has been reported, while unsaturated fat increased the quantity of *Akkermansia* and *Bifidobacterium* and reduced harmful bacteria such as *Streptococcus* and *Escherichia* spp. In addition, saturated fat can increase the *Firmicutes* to Bacteroidetes ratio and unsaturated fat can reduce the Firmicutes to Bacteroidetes ratio. Therefore, they could have varying effects on human health depending on the fat quality. Overall, numerous clinical studies suggest that a high-fat diet is deleterious for gut health, as it reduces the abundance of beneficial microbes (Yang et al. 2020).

Dietary proteins are the primary source of nitrogen for colonic microbial growth and a high intake of proteins reduces the abundance of microbes with beneficial health effects (Zhang, Ju, and Zuo 2018). It has also been implied that there was a relation between high intake of animal proteins and increased risk of IBD (Salonen and de Vos 2014). According to (Kostovcikova et al. 2019), the intake of animal-based protein increased the sensitivity to intestinal inflammation by increasing the genera of *Enterococcus*, *Streptococcus*, *Turicibater*, and *Escherichia*, and families *Peptostreptococcaceae* and *Ruminococcaceaea* in mice in comparison with a plant-based protein.

4. Effect of Gut Microbiota on GI Tract Diseases

Gut microbiota is involved in a synbiotic relationship with its host and is essential for its homeostasis, physiology, metabolism and immune functions (Stephens, Arhire, and Covasa 2018). The gut microbiota functions are categorized into metabolic, protective, structural and neurological (Adak and Khan 2019). Figure 2.1 illustrates different functionalities of the gut microbiota.

Since the gut microbiota harbors the GI tract, it plays a vital role in regulating intestinal diseases. Dysbiosis of microbiota results in various gastrointestinal disorders such as irritable bowel syndrome (IBS), IBDs and colorectal cancer (Guinane and Cotter 2013).

IBS is a gastrointestinal disorder characterized by recurrent abdominal pain or discomfort and changed bowel habits (Chey, Kurlander, and Eswaran 2015). The diagnosis of IBS is based mainly on assessing the symptoms as described by Rome criteria (El-Salhy, Hatlebakk, and Hausken 2019). Based on stool patterns, IBS patients are divided into four subtypes: diarrhea-predominant (IBS-D), constipation-predominant (IBS-C), mixed diarrhea and constipation (IBS-M), and patients who meet the diagnostic criteria for IBS, but whose bowel habits cannot be accurately categorized (IBS-U) (Spiller et al. 2007).

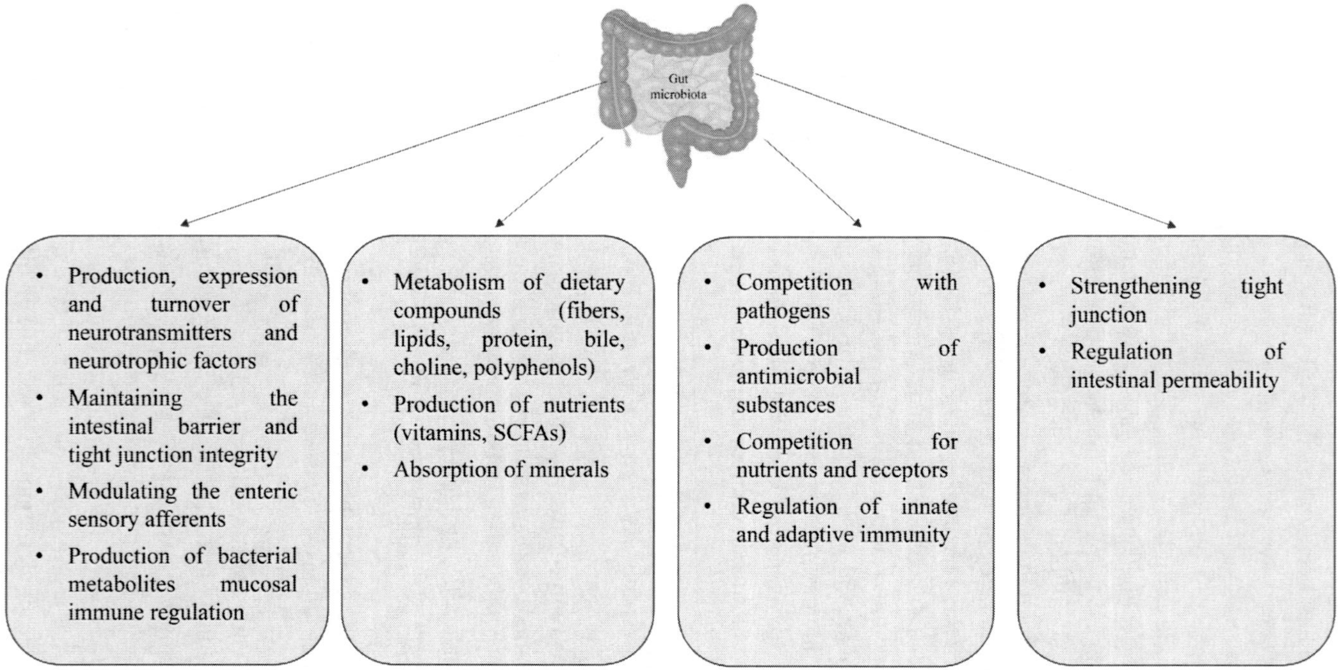

Figure 2.1. Main functions of the gut microbiota.

Several studies have reported the alteration of gut microbiota in patients with IBS compared to healthy individuals. In a study by (Rajilić–Stojanović et al. 2011), an increase in the number of *Firmicutes* including *Rumino-coccus, Clostridium* and *Dorea* along with a decrease in *Bifidobacterium* and *Faecalibacterium* spp. has been reported. According to (Chong et al. 2019; Pozuelo et al. 2015), the abundance of *Erysipelotri-chaceae* and *Ruminococcaceae* as butyrate-producing bacteria were lower in IBS patients compared to healthy ones while the population of methane-producing bacteria, *Methanobacterialles*, was higher IBS-C than in healthy subjects.

Furthermore, a higher abundance of bacteria belonging to Proteobac-teria, *Veillonella*, and Firmicutes, such as *Lactobacillus* and *Ruminococcus* and fewer *Bifidobacterium*, *Faecalibacterium*, *Erysipelo-trichaceae* and methanogens have been reported compared to healthy individuals. It has also been hypothesized that changing the intestinal permeability as a consequence of activation of the mucosal adaptive and innate immune response, increased expression of colonic Toll-like receptors (TLRs), dysregulation of the enteric nervous system and alteration of epithelial defense mechanisms that are associated with the etiology of IBS (Aziz et al. 2013; El-Salhy, Hatlebakk, and Hausken 2019).

IBD consisting of Crohn's disease (CD) and ulcerative colitis (UC) is characterized by chronic and relapsing inflammatory disorders of the gastrointestinal tract (Zuo and Ng 2018). Changes in gut microbiota diversity and decrease of Firmicutes abundance have been reported in IBD patients. Alterations in the abundance of bacterial species in IBD patients have been connected with the changes in microbiota functions such as decreased metabolism of short-chain fatty acids, predominantly acetate, propionate and butyrate, the decreased biosynthesis of amino acids and the increased auxotrophy, oxidative stress and secretion of toxins (LeBlanc et al. 2017). In addition, dysregulation of T-cells differentiation and intestinal epithelial cell death mechanisms, disruption of epithelial barrier integrity, and induction of oxidative stress are also associated with the pathogenesis of IBD (Aggeletopoulou et al. 2019).

Colorectal cancer (CRC) is one of the most common malignancies globally, with the second position in mortality and third in incidence (Bray et al. 2018). Multiple studies have indicated the role of microbiota in the pathogenesis of CRC. It has been highlighted that bacterial species such as *Clostridium septicum*, *Enterococcus faecalis*, *Streptococcus bovis*, *Bacteroi-des fragilis*, *Helicobacter pylori*, *Escherichia coli* and *Fusobacterium* spp. are involved in CRC pathogenesis (Jahani-Sherafat et al. 2018). According to

(Sobhani et al. 2011), In CRC patients, *Ruminococcus bromii*, *Clostridium clostridioforme* and *Bifidobacterium longum* have low abundance compared to normal population. Furthermore, an increase in the population of the *Bacteroides-Prevotella* and *Fusobacterium nucleatum* has been reported in CRC patients (Ray 2011). The role of microbial species has been described in several review articles (Gao et al. 2017; Tilg et al. 2018; Wong and Yu 2019). Gut microbiota can contribute to the development of CRC through induction of chronic inflammatory state or immune response, changing stem cell dynamics, the biosynthesis of toxic and genotoxic metabolites, and influencing the host metabolism (Zou, Fang, and Lee 2018).

Conclusion

The microorganisms present in the gastrointestinal tract play a substantial role in the modulation of the human metabolism and immunity. In recent years, the composition and functions of this microbial community have been established. It has been revealed that some microbes including, species from *Bifidobacterium*, *Lactobacillus*, *Akkermansia*, *Fecalibacterium*, *Eubacterium*, *Roseburia*, *Ruminococcus*, and *Blautia* are associated with a healthy gut and several factors can influence the balance between the abundance of these species with detrimental microbes known as dysbiosis. Although several studies have presented the association between gut microbiota dysbiosis and various diseases, the exact underlying mechanisms are unclear and need further investigation. However, recent researchers are moving forward to find efficient strategies for the modulation of gut microbiota, which present promising ways to prevent and manage several diseases.

References

Adak, A., and Khan, M. R. 2019. "An insight into gut microbiota and its functionalities." *Cellular and Molecular Life Sciences* 76 (3): 473-493.

Aggeletopoulou, I., Konstantakis, C., Assimakopoulos, S. F., and Triantos, C. 2019. "The role of the gut microbiota in the treatment of inflammatory bowel diseases." *Microbial Pathogenesis* 137: 103774.

Aziz, Q., Doré, J., Emmanuel, A., Guarner, F., and Quigley, E. 2013. "Gut microbiota and gastrointestinal health: current concepts and future directions." *Neurogastroenterology & Motility* 25 (1): 4-15.

Bray, F., Ferlay, J., Soerjomataram, I., Siegel, R. L., Torre, L. A., and Jemal, A. 2018. "Global cancer statistics 2018: GLOBOCAN estimates of incidence and mortality worldwide for 36 cancers in 185 countries." *CA: A Cancer Journal for Clinicians* 68 (6): 394-424.

Canfora, E. E., Jocken, J. W., and Blaak, E. E. 2015. "Short-chain fatty acids in control of body weight and insulin sensitivity." *Nature Reviews Endocrinology* 11 (10): 577.

Chey, W. D., Kurlander, J., and Eswaran, S. 2015. "Irritable bowel syndrome: a clinical review." *Jama* 313 (9): 949-958.

Chiaro, T. R., Soto, R., Stephens, W. Z., Kubinak, J. L., Petersen, C., Gogokhia, L., Bell, R., Delgado, J. C., Cox, J., and Voth, W. 2017. "A member of the gut mycobiota modulates host purine metabolism exacerbating colitis in mice." *Science Translational Medicine* 9 (380).

Chong, P. P., Chin, V. K., Looi, C. Y., Wong, W. F., Madhavan, P., and Yong, V. C. 2019. "The microbiome and irritable bowel syndrome–a review on the pathophysiology, current research and future therapy." *Frontiers in Microbiology* 10: 1136.

Danneskiold-Samsøe, N. B., Barros, H. D. d. F. Q., Santos, R., Bicas, J. L., Cazarin, C. B. B., Madsen, L., Kristiansen, K., Pastore, G. M., Brix, S., and Júnior, M. R. M. 2019. "Interplay between food and gut microbiota in health and disease." *Food Research International* 115: 23-31.

Dieterich, W., Schink, M., and Zopf, Y. 2018. "Microbiota in the gastrointestinal tract." *Medical Sciences* 6 (4): 116.

Dieterich, W., Schuppan, D., Schink, M., Schwappacher, R., Wirtz, S., Agaimy, A., Neurath, M. F., and Zopf, Y. 2019. "Influence of low FODMAP and gluten-free diets on disease activity and intestinal microbiota in patients with non-celiac gluten sensitivity." *Clinical Nutrition* 38 (2): 697-707.

Dominguez-Bello, M. G., Costello, E. K., Contreras, M., Magris, M., Hidalgo, G., Fierer, N., and Knight, R. 2010. "Delivery mode shapes the acquisition and structure of the initial microbiota across multiple body habitats in newborns." *Proceedings of the National Academy of Sciences* 107 (26): 11971-11975.

El-Salhy, M., Hatlebakk, J. G., and Hausken, T. 2019. "Diet in irritable bowel syndrome (IBS): interaction with gut microbiota and gut hormones." *Nutrients* 11 (8): 1824.

El Aidy, S., Van Den Bogert, B., and Kleerebezem, M. 2015. "The small intestine microbiota, nutritional modulation and relevance for health." *Current Opinion in Biotechnology* 32: 14-20.

Forbes, J. D., Van Domselaar, G., and Bernstein, C. N. 2016. "The gut microbiota in immune-mediated inflammatory diseases." *Frontiers in Microbiology* 7: 1081.

Gao, R., Gao, Z., Huang, L., and Qin, H. 2017. "Gut microbiota and colorectal cancer." *European Journal of Clinical Microbiology & Infectious Diseases* 36 (5): 757-769.

Ghoshal, U. C., and Ghoshal, U. 2017. "Small intestinal bacterial overgrowth and other intestinal disorders." *Gastroenterology Clinics* 46 (1): 103-120.

Guinane, C. M., and Cotter, P. D. 2013. "Role of the gut microbiota in health and chronic gastrointestinal disease: understanding a hidden metabolic organ." *Therapeutic Advances in Gastroenterology* 6 (4): 295-308.

Hollister, E. B., Gao, C., and Versalovic, J. 2014. "Compositional and functional features of the gastrointestinal microbiome and their effects on human health." *Gastroenterology* 146 (6): 1449-1458.

Holscher, H. D., Caporaso, J. G., Hooda, S., Brulc, J. M., Fahey Jr, G. C., and Swanson, K. S. 2015. "Fiber supplementation influences phylogenetic structure and functional capacity of the human intestinal microbiome: follow-up of a randomized controlled trial." *The American Journal of Clinical Nutrition* 101 (1): 55-64.

Jahani-Sherafat, S., Alebouyeh, M., Moghim, S., Amoli, H. A., and Ghasemian-Safaei, H. 2018. "Role of gut microbiota in the pathogenesis of colorectal cancer; a review article." *Gastroenterology and Hepatology from Bed to Bench* 11 (2): 101-109.

Koh, A., De Vadder, F., Kovatcheva-Datchary, P., and Bäckhed, F. 2016. "From dietary fiber to host physiology: short-chain fatty acids as key bacterial metabolites." *Cell* 165 (6): 1332-1345.

Kostovcikova, K., Coufal, S., Galanova, N., Fajstova, A., Hudcovic, T., Kostovcik, M., Prochazkova, P., Jiraskova Zakostelska, Z., Cermakova, M., and Sediva, B. 2019. "Diet rich in animal protein promotes pro-inflammatory macrophage response and exacerbates colitis in mice." *Frontiers in Immunology* 10: 919.

LeBlanc, J. G., Chain, F., Martín, R., Bermúdez-Humarán, L. G., Courau, S., and Langella, P. 2017. "Beneficial effects on host energy metabolism of short-chain fatty acids and vitamins produced by commensal and probiotic bacteria." *Microbial Cell Factories* 16 (1): 1-10.

Liu, H., Wang, J., He, T., Becker, S., Zhang, G., Li, D., and Ma, X. 2018. "Butyrate: a double-edged sword for health?" *Advances in Nutrition* 9 (1): 21-29.

Manrique, P., Bolduc, B., Walk, S. T., van der Oost, J., de Vos, W. M., and Young, M. J. 2016. "Healthy human gut phageome." *Proceedings of the National Academy of Sciences* 113 (37): 10400-10405.

Mariat, D., Firmesse, O., Levenez, F., Guimarães, V., Sokol, H., Doré, J., Corthier, G., and Furet, J. 2009. "The Firmicutes/Bacteroidetes ratio of the human microbiota changes with age." *BMC Microbiology* 9 (1): 1-6.

Morrison, D. J., and Preston, T. 2016. "Formation of short chain fatty acids by the gut microbiota and their impact on human metabolism." *Gut Microbes* 7 (3): 189-200.

Norman, J. M., Handley, S. A., Baldridge, M. T., Droit, L., Liu, C. Y., Keller, B. C., Kambal, A., Monaco, C. L., Zhao, G., and Fleshner, P. 2015. "Disease-specific alterations in the enteric virome in inflammatory bowel disease." *Cell* 160 (3): 447-460.

Palmer, C., Bik, E. M., DiGiulio, D. B., Relman, D. A., and Brown, P. O. 2007. "Development of the human infant intestinal microbiota." *PLoS Biol* 5 (7): e177.

Parkar, S. G., Stevenson, D. E., and Skinner, M. A. 2008. "The potential influence of fruit polyphenols on colonic microflora and human gut health." *International Journal of Food Microbiology* 124 (3): 295-298.

Pozuelo, M., Panda, S., Santiago, A., Mendez, S., Accarino, A., Santos, J., Guarner, F., Azpiroz, F., and Manichanh, C. 2015. "Reduction of butyrate-and methane-producing microorganisms in patients with Irritable Bowel Syndrome." *Scientific Reports* 5 (1): 1-12.

Rajilić–Stojanović, M., Biagi, E., Heilig, H. G., Kajander, K., Kekkonen, R. A., Tims, S., and de Vos, W. M. 2011. "Global and deep molecular analysis of microbiota signatures in fecal samples from patients with irritable bowel syndrome." *Gastroenterology* 141 (5): 1792-1801.

Rajoka, M. S. R., Shi, J., Mehwish, H. M., Zhu, J., Li, Q., Shao, D., Huang, Q., and Yang, H. 2017. "Interaction between diet composition and gut microbiota and its impact on gastrointestinal tract health." *Food Science and Human Wellness* 6 (3): 121-130.

Ray, K. 2011. "Fusobacterium nucleatum found in colon cancer tissue—could an infection cause colorectal cancer?" *Nature Reviews Gastroenterology & Hepatology* 8 (12): 662-662.

Rodríguez, J. M., Murphy, K., Stanton, C., Ross, R. P., Kober, O. I., Juge, N., Avershina, E., Rudi, K., Narbad, A., and Jenmalm, M. C. 2015. "The composition of the gut microbiota throughout life, with an emphasis on early life." *Microbial Ecology in Health and Disease* 26: 26050.

Salonen, A., and de Vos, W. M. 2014. "Impact of diet on human intestinal microbiota and health." *Annual Review of Food Science and Technology* 5: 239-262.

Santino, A., Scarano, A., De Santis, S., De Benedictis, M., Giovinazzo, G., and Chieppa, M. 2017. "Gut microbiota modulation and anti-inflammatory properties of dietary polyphenols in IBD: New and consolidated perspectives." *Current Pharmaceutical Design* 23 (16): 2344-2351.

Sekirov, I., Russell, S. L., Antunes, L. C. M., and Finlay, B. B. 2010. "Gut microbiota in health and disease." *Physiological Reviews* 90: 859-904.

Selma, M. V., Espin, J. C., and Tomas-Barberan, F. A. 2009. "Interaction between phenolics and gut microbiota: role in human health." *Journal of Agricultural and Food Chemistry* 57 (15): 6485-6501.

Sender, R., Fuchs, S., and Milo, R. 2016. "Are we really vastly outnumbered? Revisiting the ratio of bacterial to host cells in humans." *Cell* 164 (3): 337-340.

Shukla, S. D., Budden, K. F., Neal, R., and Hansbro, P. M. 2017. "Microbiome effects on immunity, health and disease in the lung." *Clinical & Translational Immunology* 6 (3): e133.

Sobhani, I., Tap, J., Roudot-Thoraval, F., Roperch, J. P., Letulle, S., Langella, P., Corthier, G., Van Nhieu, J. T., and Furet, J. P. 2011. "Microbial dysbiosis in colorectal cancer (CRC) patients." *PloS one* 6 (1): e16393.

Sokol, H., Leducq, V., Aschard, H., Pham, H.-P., Jegou, S., Landman, C., Cohen, D., Liguori, G., Bourrier, A., and Nion-Larmurier, I. 2017. "Fungal microbiota dysbiosis in IBD." *Gut* 66 (6): 1039-1048.

Spiller, R., Aziz, Q., Creed, F., Emmanuel, A., Houghton, L., Hungin, P., Jones, R., Kumar, D., Rubin, G., and Trudgill, N. 2007. "Guidelines on the irritable bowel syndrome: mechanisms and practical management." *Gut* 56 (12): 1770-1798.

Stephens, R. W., Arhire, L., and Covasa, M. 2018. "Gut microbiota: from microorganisms to metabolic organ influencing obesity." *Obesity* 26 (5): 801-809.

Sun, H., Chen, Y., Cheng, M., Zhang, X., Zheng, X., and Zhang, Z. 2018. "The modulatory effect of polyphenols from green tea, oolong tea and black tea on human intestinal microbiota in vitro." *Journal of Food Science and Technology* 55 (1): 399-407.

Thursby, E., and Juge, N. 2017. "Introduction to the human gut microbiota." *Biochemical Journal* 474 (11): 1823-1836.

Tilg, H., Adolph, T. E., Gerner, R. R., and Moschen, A. R. 2018. "The intestinal microbiota in colorectal cancer." *Cancer Cell* 33 (6): 954-964.

Wong, S. H., and Yu, J. 2019. "Gut microbiota in colorectal cancer: mechanisms of action and clinical applications." *Nature Reviews Gastroenterology & Hepatology* 16 (11): 690-704.

Yang, Q., Liang, Q., Balakrishnan, B., Belobrajdic, D. P., Feng, Q.-J., and Zhang, W. 2020. "Role of dietary nutrients in the modulation of gut microbiota: a narrative review." *Nutrients* 12: 381.

Zhang, N., Ju, Z., and Zuo, T. 2018. "Time for food: The impact of diet on gut microbiota and human health." *Nutrition* 51: 80-85.

Zhao, L., Zhang, Q., Ma, W., Tian, F., Shen, H., and Zhou, M. 2017. "A combination of quercetin and resveratrol reduces obesity in high-fat diet-fed rats by modulation of gut microbiota." *Food & Function* 8 (12): 4644-4656.

Zou, S., Fang, L., and Lee, M.-H. 2018. "Dysbiosis of gut microbiota in promoting the development of colorectal cancer." *Gastroenterology Report* 6 (1): 1-12.

Zuo, T., and Ng, S. C. 2018. "The gut microbiota in the pathogenesis and therapeutics of inflammatory bowel disease." *Frontiers in Microbiology* 9: 2247.

Chapter 3

Bioactive Compounds in Foods: History, Definitions and Classifications

Abstract

Bioactive compounds are substances present in small amounts in foods and provide beneficial health effects. These compounds vary extensively in chemical structure and function and are categorized into different groups, including phenolic compounds, carotenoids, phytoestrogens, phytosterols and phytostanols, probiotics, prebiotics, as well as bioactive lipids, carbohydrates and peptides. It has been revealed that these bioactive compounds can have neuroprotective, cardioprotective, hepatoprotective, antioxidant, antimicrobial, antidiabetes, antiobesity and anticancer effects through several mechanisms. In this chapter, the structure and classification of bioactive compounds and their potential health benefits are described.

Keywords: bioactivity, chronic disease, functional food, health effect

1. Introduction

In 1989, Dr. Stephen L. DeFelice introduced the term nutraceutical and defined it as 'any substance that is a food or part of a food that provides medical and/or health benefits, including the prevention and treatment of disease' (DeFelice 2002). This definition covers foods, dietary supplements, and medical foods. Nutraceuticals are defined in the Nutraceuticals Institute (Rutgers and St. Joseph Universities) as follows: 'Nutraceuticals are natural bioactive chemical compounds with health-promoting, disease-preventing, and medicinal properties which are usually referred to as phytochemicals or functional foods (Anonymous. 2002).

The terms 'nutraceutical' and 'functional food' are often used interchangeably, albeit the latter has recently become more common. By definition, foods containing bioactive compounds are called functional foods (Shahidi 2009). Bioactive compounds deliver physiological benefits related to health promotion and disease prevention (Aluko 2011; Awika 2011). Functional foods provide essential nutrients necessary for health, but they can also prevent diseases caused by nutritional deficiencies (Abdul-Hamid and Luan 2000).

Part of the beneficial effects of functional foods is due to their antioxidant properties. It should be considered that functional foods might have significant antioxidant activity within the gastrointestinal tract, even if not absorbed (Jovanovic and Simic 2000). These compounds can modulate cellular biochemical pathways, especially in cases such as cancer that some of these pathways are hyperactive. Moreover, functional food components can regulate gene expression; for example, it has been revealed that these ingredients can suppress some of the genes that are upregulated in inflammation and cancer, including tumor necrosis factor-alpha (TNF-α), interleukins and protein kinase C (Paliyath and Shetty 2011).

The use of herbs, plant extracts, or chemicals derived from plants for medical purposes has a history of several years. Interest in the medicinal properties of herbs is on the rise universally; hence, today, many pharmacological drugs are also produced based on plant compounds such as oils and extracts. The use of medicinal plants and natural compounds, in addition to cost savings, has fewer side effects (Gilani 2005). The following sections describe various bioactive compounds and their beneficial effects.

2. Phenolic Compounds

Polyphenols and other phenolic compounds are the most broadly distributed class of plant secondary compounds. An essential feature of polyphenols is the presence of several phenolic rings along with two or more hydroxyl substitutions. Substituting the basic chemical structure of polyphenols, polymerization, and oxidation can lead to a wide variety of polyphenol chemical structures. Accordingly, polyphenols are classified as flavonoids and non-flavonoids, comprising several subgroups. Flavonoid-type polyphenols can be further divided into classes based on the oxidation state of the central pyran ring and mainly includes anthocyanins, flavones, catechins, isoflavones, and flavonols. Compounds such as stilbenes, phenolic acids and lignans are classified in the non-flavonoid group (Daglia 2012; Tsao 2010; Farzaei, Rahimi, and Abdollahi 2015; Edwards-Jones et al. 2008).

Since polyphenols are ubiquitously present in fruits and vegetables, they are considered an integral part of the human diet. Polyphenols are the largest group of phytochemicals and the most abundant dietary antioxidants. In addition to fruits and vegetables, olive oil, chocolate, and beverages such as coffee, tea, and wine are among the richest sources of polyphenols (Lee, Koo, and Min 2004). Their content is indexed in nutrient databanks (Claudine Manach et al. 2004). It should be noted that fruits, vegetables and other sources of polyphenol contain multiple polyphenol classes and other nutritious compounds (Hilz et al. 2005; Prior and Gu 2005). For instance, although cocoa, berries and grapeseed are rich in proanthocyanidins, they also encompass other bioactive components, including catechins and anthocyanins (Martin and Bolling 2015).

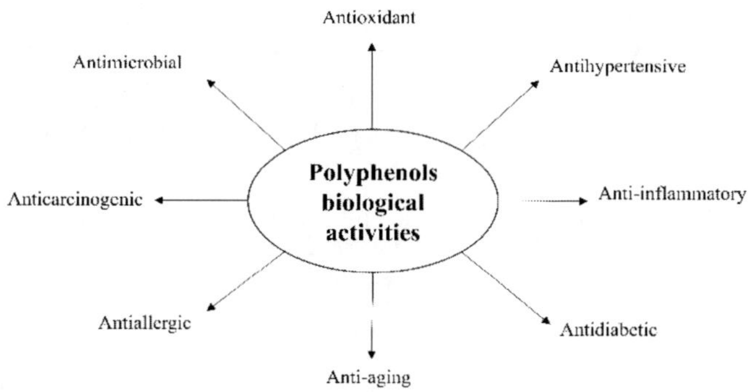

Figure 3.1. Beneficial effects of polyphenols.

A balanced diet provides around 1 gram of polyphenol daily (Scalbert and Williamson 2000). Dietary intake of polyphenols in individuals who consume several servings of fruits, vegetables, tea, or coffee can reach several grams per day. For example, habitual coffee drinking can provide up to 800 mg of hydroxycinnamic acids per day (Claudine Manach et al. 2004).

As presented in Figure 3.1, numerous healthful properties are mentioned for polyphenols. Some epidemiological studies have stated an inverse relationship between the consumption of phenolic compounds and the risk of certain diseases such as cancer and cardiovascular diseases (Lee, Koo, and Min 2004; Magrone et al. 2008). Due to their anti-inflammatory and antioxidant properties, polyphenols can modulate cellular signaling pathways (Edwards-Jones et al. 2008). Polyphenols also have an antimicrobial activity that can be effective in gut microbial balance regulation. Moreover, many phenolic compounds are used in the pharmaceutical field, such as ellagic acid, caffeic acid, quercetin, gingerol, kaempferol, resveratrol and cyanidin (Antolovich et al. 2000; Daglia 2012). The high amounts of some polyphenols, such as tannin, can be nutritionally undesirable. Chelating and reducing proteins digestibility, hepatotoxicity, and increased risk of cancer with high levels of some polyphenols have been reported, while small quantities may be beneficial (Chung, Wei, and Johnson 1998).

Table 3.1. Selected publications on the incorporation of polyphenols into food products

Food product	Source of polyphenol	Reference
Low-fat hard cheese	Green tea catechin and epigallocatechin gallate within soy lecithin liposomes	(Rashidinejad et al. 2014)
Yogurt ice-cream	Apple peel extract (1-5%)	(Ahmad et al. 2020)
Stirred yogurt	Cherry extract in chitosan-coated liposomes (5% w/v)	(Akgün et al. 2020)
Cheese	Grape pomace powder (1% w/w)	(Gaglio et al. 2021)
Petit-Suisse goat cheese	Acerola (*Malpighia emarginata* DC) pulp (10%)	(de Barcelos et al. 2020)
"Sardaigne" cheese	*Arbutus unedo* L. extract	(Masmoudi et al. 2020)
Cottage cheese	Rosemary extract (0.1 g of free and 0.9 g of alginate-base microencapsulated in 200 g of cheese)	(Ribeiro et al. 2016)
Cheddar chees	Green tea extract (0.1% w/w)	(Lamothe et al. 2016)
Yogurt	Anthocyanin-rich riceberry rice (0.125–0.25% (w/w)	(Anuyahong, Chusak, and Adisakwattana 2020)
Yogurt	Encapsulated Hibiscus Extract (20% w/w)	(de Moura et al. 2019)
Chocolate vanilla dairy drink	Encapsulated Arjuna herb (0.3%)	(Sawale et al. 2017)
Cookie	Peach puree (10.5%)	(Canalis et al. 2020)
Bread	Vegetables leaf (*Amaranthus viridis*, *Solanum macrocarpon*, and *Telfairia occidentalis*) (3%)	(Alashi et al. 2019)
Bread	Orange peel powder (7%)	(L. Han, Zhang, and Cao 2021)
Muffins and cookies	Goji berry by-product (10-40% w/w)	(Bora, Ragaee, and Abdel-Aal 2019)
Cake	Red onion peel (3%)	(Elsebaie and Essa 2018)
Cookie	Encapsulated sour cherry pomace extract (10 and 15%)	(Šaponjac et al. 2016)
Biscuit	Bamboo shoot powder (5, 10, 15%)	(Choudhury et al. 2015)
Semi-sweet biscuit	Sea grapes (Caulerpa racemosa) (1, 5, 10%)	(Kumar et al. 2018)
Pasta	Grape marc	(Marinelli et al. 2015)
Pasta	Syzygium cumini L. pulp (10-40%)	(Panghal et al. 2019)
Pasta	Black mulberry extract (0.5, 1, 1.5%)	(Yazdankhah, Hojjati, and Azizi 2019)
Cantonese sausages	Mulberry (0.5 or 1.0 g/kg)	(Xiang et al. 2019)
Sausage	Flaxseed and tomato powder (3, 6%)	(Ghafouri-Oskuei et al. 2020)
Fish sausage	Onion peel powder (1-3%)	(Bedrníček et al. 2020)
Pork sausage	Green and black tea (0.05%, 0.10, 0.20%, and 0.30%)	(Jayawardana et al. 2019)

The beneficial functions of polyphenols have led to an increase in their consumption and production. The amount of industrially extracted polyphenols reached 16,380 tons in 2015 and is expected to increase to 33,880 tons by the end of 2024. The industrial utilization of polyphenols is dominated by functional beverages (44%) and food (33%) (Adebooye, Alashi, and Aluko 2018). In recent years, polyphenols or polyphenol-enriched by-products have been extensively incorporated into several food products to develop functional foods. Table 3.1 presents some food products enriched by polyphenols from different sources.

2.1. Stilbenes

Stilbenes are phytoalexins synthesized as a part of the plant's defensive mechanism in response to pathogenic infection (Figure 3.2). Resveratrol is a natural dietary phenolic compound and the most famous member of stilbenes which is produced in grape skin cells (Creasy and Coffee 1988; Jeandet, Bessis, and Gautheron 1991); however, it is also found in other food items like berries and peanuts (Yadav et al. 2009). The therapeutic properties of resveratrol, including anti-carcinogenic, chemopreventive, and anti-aging effects, have led to its extensive application in the treatment of cardiovascular diseases, cancer, and neurodegenerative diseases (Tomé-Carneiro et al. 2013). Most researchers have elucidated the health beneficial effects of resveratrol, including reducing insulin resistance and glucose intolerance, modifying plasma lipid profiles, and suppression of inflammation and oxidative stress (Frémont 2000). Since resveratrol has a rapid metabolism and elimination, many studies have been executed to improve resveratrol bioavailability. Strategies proposed to overcome poor bioavailability of resveratrol are largely based on using cytochrome P450 (CYPs) inhibitors such as piperine, modifying the structure of resveratrol, or encapsulation (Santos, Veiga, and Ribeiro 2011).

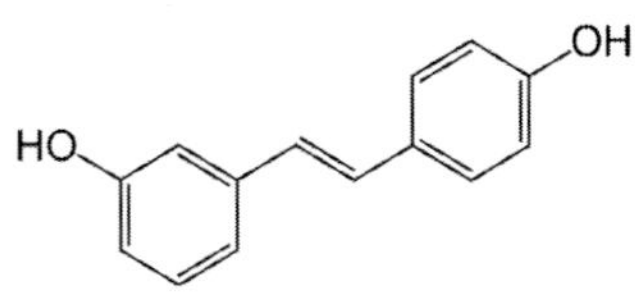

Figure 3.2. Structure of stilbenoid.

2.2. Flavonoids

The coloring of plants is the most critical function of flavonoids. For instance, anthocyanins bring red and purple color and other flavonoid compounds give white or yellow color to plants (S. Jovanovic et al. 1998).

Flavonoids are found in various forms of glycosylated, esterified, or polymerized in the diet. The effect of different flavonoids in the body is diverse due to the variances concerning absorption, metabolism and excretion. Also, in general, flavonoids have poor bioavailability and most of the dietary flavonoids remain unabsorbed in the gastrointestinal tract (Skinner and Hunter 2013). Perhaps this is why flavonoids have directly significant effects on the gastrointestinal tract, including reducing prooxidant iron and reactive oxygen species levels and possibly inhibition of cyclooxygenases and lipoxygenases (Halliwell, Rafter, and Jenner 2005). In addition, flavonoids re-enter the gastrointestinal tract through the biliary circulation. These non-absorbed compounds eventually move into the colon, where they are affected by gut microflora alteration and metabolites such as phenolic acids are produced (Del Rio et al. 2010; Roowi et al. 2009; Williamson and Clifford 2010). A 'second wave' of absorption has caused many colonic metabolites of unabsorbed flavonoids to be detectable in plasma (Del Rio et al. 2010; Roowi et al. 2009). Dietary flavonol quercetin, flavanol catechin, flavanone hesperidin, and daidzein isoflavone are some of the most common flavonoids (Figure 3.3) (Boots, Haenen, and Bast 2008).

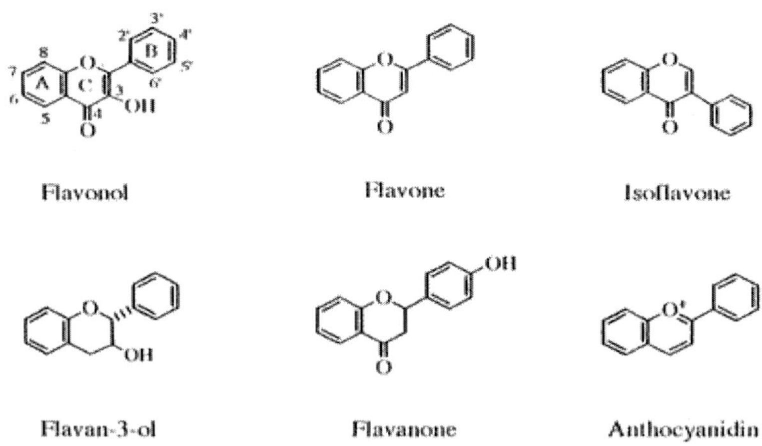

Figure 3.3. Some of the most common structures of flavonoids.

The biological properties of flavonoids are not limited to their antioxidant effects (Depeint et al. 2002). According to epidemiological studies, flavonoids have broad biological activities and scavenge free radicals (Wach, Pyrzyńska, and Biesaga 2007; Havsteen 2002). One of the pharmacological activities of flavonoids is suppression of the nuclear factor-κB (NF-κB) and activator protein 1 (AP-1) activation pathways (Piao et al. 2013; Youn et al. 2013; X. Chen et al. 2012). Flavonoids can act as antiulcer agents; they can also reduce intestinal motility and secretions and in light of this, flavonoids can be considered anti-diarrheal remedies (Lima et al. 2005; Carlo et al. 1993; La Casa et al. 2000).

2.3. Flavonols

Quercetin, a natural aglycone polyphenol belonging to a flavonoids family, is one of the most prominent dietary antioxidants (Yoshino et al. 2012). It is present in many fruits and vegetables, mainly found in onions, broccoli, tomatoes, apple peel, pomegranate and berries. Daily dietary intake of quercetin in a balanced diet has been assessed around 5-40 mg (Yoshino et al. 2012). Subsequently, quercetin plasma levels reached less than 100 nmol/L. This increase was higher with dietary supplements; therefore 28 days of supplementation with one gram of quercetin per day could increase the plasma concentrations to 1.5 μmol/L. It is worth mentioning that not only glycosides in food sources affect quercetin availability, but gut microflora can also alter the absorption of this polyphenol (Claudine Manach et al. 2005; Conquer et al. 1998). Quercetin has an extensive metabolism in the body and the post-absorption plasma concentration of this flavonol is negligible at the low nanomolar range (Skinner and Hunter 2013).

Part of the beneficial effects of quercetin is due to its ability to eliminate reactive species such as free radicals (Boots, Haenen, and Bast 2008) and its antioxidant effect is even greater than vitamin E and C. Quercetin has protective effects against cardiovascular disease, atherosclerosis, hypertension and diabetes. It may also be useful in the treatment of cancer due to its anti-proliferative and anti-metastatic effects (Russo et al. 2012).

2.4. Flavanones

Flavanones with different glycosides of three major aglycones including hesperidin, naringenin and eriodictyol, constitute a unique subgroup of flavonoids (Manach et al. 2003). In contrast to other flavonoids, flavanones are not widely distributed in foods and are found almost exclusively in citrus fruits and, to a lesser extent, in tomatoes and some aromatic herbs (Chanet et al. 2012; Peterson et al. 2006). However, based on findings of a recent paper on polyphenol intake in Finland, hesperidin appears to be the major consumed flavonoid (Knekt et al. 2002).

As previously mentioned, flavonoids have broad biological properties, including antioxidant and anti-inflammatory effects, so it is not surprising that hesperidin also exhibits these properties (Harborne and Williams 2000; Pietta 2000). Hesperidin is called bioflavonoid due to its various biological activity. The ability to alter the gene expression is another feature of hesperidin toward an anti-inflammatory and atherogenic profile (Milenkovic et al. 2011). A conducted trial on individuals with metabolic syndrome revealed that supplementation with 500 mg hesperidin could significantly reduce plasma levels of two inflammatory biomarkers, C-reactive protein (CRP) and serum amyloid A (SAA) (Rizza et al. 2011). In addition to antioxidative and anti-diabetic activity, hesperidin can trap reactive radicals (Hirata et al. 2005). From this perspective, hesperidin can be an effective remedy in the fatty liver as it inhibits both the synthesis and absorption of cholesterol by changing the expression of mRNA for some of the proteins involved in this pathway (Wang et al. 2011). Reducing the gene expression of cyclooxygenase-2 (COX-2) (Hirata et al. 2005; Emim, Oliveira, and Lapa 1994), modifying serum lipid profiles (Monforte et al. 1995), improving glucose and lipid metabolism (Miwa et al. 2005; Miwa et al. 2004), and suppressing lipid peroxidation through activated receptors (PPARs) are other reported effects for hesperidin (Agrawal et al. 2014). All of the mentioned effects are part of the anti-inflammatory activity of hesperidin.

2.5. Flavanols

Flavanol is also called flavan- 3-ol, which refers to hydroxyl group in the 3-position (Chung et al. 1998). Although flavan-3-ols, particularly (+)-catechin and (−)-epicatechin are widely distributed in plants, they often occur in chocolate, grapes and berries (Rothwell et al. 2013; Shahidi and Naczk 1995).

Green tea also represents a rich source of monomeric flavanols, mainly four major catechins, including epicatechin, epicatechin-3 -gallate, epigallocatechin (EGC), and epigallocatechin-3-gallate (EGCG), which are principal polyphenols in terms of medicinal effects (Graham 1992; Blumberg et al. 2015). It is worth noting that the polyphenolic content of black tea is distinct from green tea. Another kind of tea called Oolong tea falls between green and black tea regarding the flavanol content (Watson and Preedy 2012).

Despite the significant content of flavonoids in green tea, it appears that the most abundant catechin is EGCG, which is reported to be up to 75% (Chen et al. 2001). Epigallocatechin-3-gallate is also the most excellent antioxidant and anti-inflammatory ingredient of green tea, and therefore possesses many health benefits, including anti-diabetic, anti-cancer, and cardioprotective functions (Yang et al. 2001). The high content of antioxidants in green tea can explain the established favorable effects on human metabolism (Grosso et al. 2015; Chen et al. 2001).

As it was cited in the literature, cacao is another rich source of flavanol, which mainly contains catechin and EC. The antioxidant properties of cocoa and its ability to constrain the amylase and glycosidase activity are also attributed to these components (Satoh et al. 2015; Matsuzaki and Hara 1985; Rodríguez-Ramiro et al. 2011).

2.6. Isoflavones

Isoflavones are a type of flavonoids recognized as phytoestrogen due to estrogen-like activity. Twelve types of isoflavonoid components have been identified: derivatives of daidzein, genistein, and glycitein (Jackson et al. 2002). The form in which isoflavones are present in foodstuff determines their bioavailability and bioactivity; for case, the aglycons absorption rate is greater than that of glucosides (Zheng et al. 2004; Nollet and Toldrá 2012). The main dietary source of isoflavones is leguminous plants, particularly soybean; however, they can be found in animal products such as meat, eggs, or dairy due to dietary intake (Kuhnle et al. 2008; KeShun Liu 1997; Messina 1997).

Although isoflavones have estrogen-like activity, their mechanism of action is different from estrogen (Fritz et al. 2013). Anti-oxidative activity and reduced tissue damage caused by free radicals are part of the effects considered for isoflavones (Han et al. 2009). Genistein also acts as an activator for PPAR and induces the expression of superoxide dismutase (Sueishi et al. 2014; Zhang et al. 2013). All of these members have anticancer effects towards different cancers; for instance, the positive effects of genistein, found in peas, lentils, soybeans, and other beans, in liver cancer have been revealed (Spagnuolo et al. 2015).

The weak estrogenic activity of isoflavones is a factor that can compete with the more potent endogenous estrogens (e.g., 17 β-estradiol) for the estrogen receptor or may interfere with estrogen receptor action and signal transduction processes and, therefore, imposes anti-estrogenic effects. This can explain why the risk of estrogen-dependent cancers is lower in societies that consume large quantities of soy products (Messina, Barnes, and Setchell 1997). Recent studies on animal models of liver cancer have shown that genistein can be a therapeutic option in cancer by inducing apoptosis and inhibiting cell proliferation (Spagnuolo et al. 2015). In addition, through the antihypertensive effects, genistein can effectively reduce the risk of cardiovascular diseases (Hügel et al. 2016).

2.7. Anthocyanins

Anthocyanins are water-soluble red, purple, and blue natural pigments present abundantly in fruits and vegetables such as blueberries, strawberries, cherries, plums, grapes, red oranges, purple cauliflower and red onions (Valenti et al. 2013; Afford 2000). An exciting aspect of the anthocyanin action in coloring plants is changing plant colors by changing the season (Cooper-Driver 2001). Anthocyanins exist largely in plants and are the most common class of phenolic compounds. To date, more than 500 types of anthocyanins have been identified (Sarma, Sreelakshmi, and Sharma 1997).

It has been indicated that anthocyanins have positive effects on the treatment of various diseases; in many countries, anthocyanin is prescribed as a medicine and its therapeutic effects have been clarified in improving dyslipidemia (Qin et al. 2009), enhancing insulin sensitivity (Li et al. 2015), reducing inflammatory reaction and boosting antioxidant capacity (Yong Wang et al. 2016; Hassellund et al. 2013). In addition, the beneficial effects of anthocyanin on oxidative stress reduction of hepatocytes have been shown in many in vitro studies (Grosso et al. 2013). Anthocyanin is very reactive towards reactive oxygen species (Cao et al. 2001).

2.8. Chalcones

The phenolic antioxidant, curcumin is the most prominent member of this family. Curcumin, an active ingredient and the cause of the yellow color of turmeric, is frequently used as a food spice and food coloring agent. The history of using curcumin in Ayurvedic medicine dates back to several thousand years, originating in Asia. Even at relatively high oral doses, no significant side effects are known for curcumin (Aggarwal et al. 2007; Watson and Preedy 2010).

Corresponding to other polyphenols, the most distinctive features of curcumin are its antioxidant and anti-inflammatory effects (Jurenka 2009). Other positive biological activities of curcumin include cardioprotective, hepatoprotective, anti-arthritic, hypocholesterolemic and hypolipidemic properties (Vecchi Brumatti et al. 2014).

Many investigations have revealed that curcumin has tremendous therapeutic effects for treating asthma, diabetes, acute infection and gastrointestinal diseases such as non-alcoholic fatty liver and irritable bowel syndrome (Jagetia and Aggarwal 2007; Aggarwal and Harikumar 2009). Curcumin is a potent inhibitor for nuclear factor-kappa B (NF-κB) and therefore can possess many affirmative effects, such as anti-microbial, anti-tumor, and anti-inflammatory properties (Epstein, Sanderson, and MacDonald 2010; Martins et al. 2008; Sintara et al. 2010; Rudrappa and Bais 2008).

2.9. Phenolic Acids

The derivatives of benzoic acid and cinnamic acid are the major phenolic acids of the herbs. These phenolic compounds can be found in free or bound form in a wide variety of fruits, vegetables, and bran, grain, seed coat and seed oil. The bounded type of phenolic acids will only be available through hydrolysis with acid, alkali, or enzymatic digestion. Gallic acid is a kind of benzoic acid that forms ester bonding with sugar molecules (Paliyath, Bakovic, and Shetty 2011). One of the most abundant phenolic acids is caffeic acid, mainly found in coffee, formed by the esterification of quinic acid as in chlorogenic acid (Feng et al. 2005; Cha et al. 2014; Sahin et al. 2010). These phenolic acids have a radical-trapping function, oxygen removal capacity and protective properties against oxidative damage. The ability of instant coffee to scavenge hydroxyl radical is 20 times that of blueberries (Mellor et al. 2013; Spencer et al. 2001; Paliyath, Bakovic, and Shetty 2011). Therefore, coffee consumption precludes LDL oxidation and oxidative damage to nucleic acids (Sueishi et al. 2014; Laranjinha, Almeida, and Madeira 1994; Tang and Liu 2008). The anti-diabetic effects of chlorogenic acid are attributed to its ability to inhibit glucosidase activity and consequently carbohydrate degradation and prevent postprandial elevation of blood glucose levels via inducing glucagon-like peptide 1 (GLP-1) secretion (Stodt et al. 2014; Jokura et al. 2015). The suppression of fat utilization in the body and subsequently visceral fat accumulation is another effect of chlorogenic acid-rich coffee extract (Ota et al. 2010; Nagao et al. 2009). It has also been revealed that chlorogenic acid affects lipid metabolism through activation of hepatic PPAR. Administration of chlorogenic acid to experimental animals has modified serum lipid profile and reduced blood glucose and insulin levels, which was attributed to amplified expression of PPAR in the liver (Shu-Yuan et al. 2009).

3. Carotenoids

Carotenoids are fat-soluble pigments whose biosynthetic and basic functions originate from photosynthesis (Mazza 1998). Carotenoids are extensively distributed in fruits and vegetables, giving them a yellow, red, and orange color. Of the at least 600 carotenoids known in nature, about forty of these are consumed through the diet, but only a small number are identified in the blood and tissues of the human body (Khachik, Beecher, and Smith 1995; Agarwal and Rao 2000). The major components of these lipophilic molecules are β/α - carotene, lycopene, β-cryptoxanthin, lutein and zeaxanthin (Figure 3.4) (Kaulmann and Bohn 2014; Rao and Rao 2007; Johnson 2002).

Carotenoids can be divided into two groups. The first group is called pro-vitamin A and includes β-carotene α-carotene and β-cryptoxanthin, which accumulate in the skin and give it a yellow color. The second group, non-pro-vitamin A carotenoids, comprises lycopene, lutein and zeaxanthin, which accumulate preferentially in the macula lutea, where their special biological function is to protect the retina against UV-induced oxidative damage and inflammation (Kaulmann and Bohn 2014; Scarmo et al. 2010). Generally, carotenoids are potent antioxidants that protect cells against oxidative damage caused by reactive oxygen species (ROS) (Rao and Rao 2007). However, various effects have been described for each carotenoid; for example, studies have revealed the protective effects of lycopene against atherosclerosis (Palozza et al. 2011). Moreover, lycopene can lower LDL cholesterol in hyperlipidemic patients with similar mechanisms of statins (Ried and Fakler

2011). Some carotenoids are used as supplements to fortify foods or dyes in food and beverages (Eggersdorfer and Wyss 2018). Table 3.2 shows the incorporation of some carotenoids into food products.

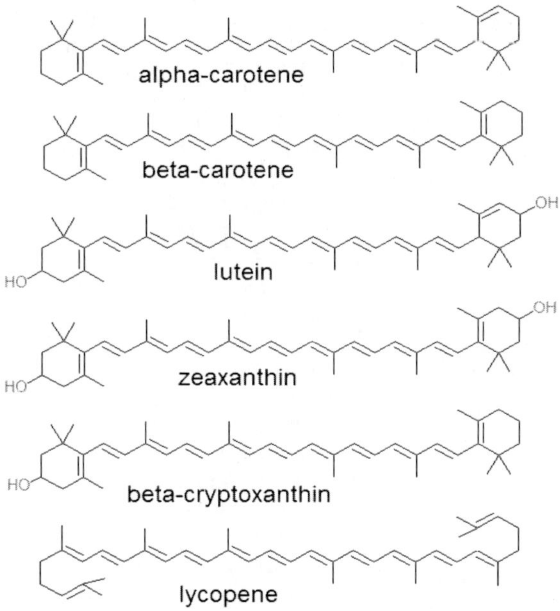

Figure 3.4. Structure of different carotenoids.

3.1. β-Carotene

β-carotene is the primary dietary source of pro-vitamin A carotenoid. β-carotene has an extensive distribution in yellow-orange and dark green fruits and vegetables (Rodriguez-Amaya et al. 2008). The most important dietary sources of β-carotene include green leafy vegetables, carrots, apricots, sweet potatoes, pumpkins and mangoes (Maiani et al. 2009; Pan and Ho 2008). β-carotene is also found to be abundant in liver (Vitaglione et al. 2005). Since β-carotene is partly converted to vitamin A, it has the most pro-vitamin activity (Krinsky and Johnson 2005). Antioxidant effects are one of the most prominent activities of β-carotene that can be mentioned through the scavenging of free radicals (Diplock 1991). Another probable effect of β-carotene is its ability to produce major *in vivo* changes in human immune cellular markers (Prabhala et al. 1991).

Table 3.2. Selected publications on the incorporation of carotenoids into food products

Food product	Type of carotenoid	Major findings	Reference
Cupcake	Lutein	Addition of water-soluble form of lutein showed satisfactory stability degree and resistance to the baking process. The recipe also showed suitable bioaccessible efficiency	(Xavier et al. 2018)
Wholegrain baked foods	Lutein	Lutein bioaccessibility of cookies and muffins was higher than flatbreads due to their elevated fat content.	(Read, Wright, and Abdel-Aal 2015)
Yogurt	Zeaxanthin	Incorporation of zeaxanthin nanoparticles into yogurt provided higher stability and promoted controlled release.	(de Campo et al. 2019)
Yogurt	β-carotene	β -carotene-loaded solid lipid microparticles stabilized with a hydrolyzed soy protein isolate incorporated into yogurt did not affect physicochemical and rheological properties of yogurt and it was approved by consumers.	(Molina et al. 2019)
Sheep milk yogurt	Lutein	In yogurt samples with high lutein content, post-acidification was observed. For all treatments, up to the end of storage, the final amount of lutein characterized the product as a nutritional source of this element.	(Barros et al. 2020)
Potato soup	Lutein	Lutein extracted from yellow corn was successfully utilized for development of ready-to-eat potato soup with minimum loss of lutein during processing.	(Pal and Bhattacharjee 2019)
Pearl Millet Based Cookies	β-carotene	Microencapsulated β-carotene improved antioxidant activity and sensory properties of cookies. The physical and textural characteristics of cookies were slightly affected by MBC supplementation.	(Kulthe, Thorat, and Lande 2017)
Synbiotic food with inulin and *Lactobacillus sporogenes*	β-carotene	β-carotene fortified synbiotic food intake in patients with type 2 diabetes for 6 weeks had favorable effects on insulin metabolism, triglycerides, VLDL-cholesterol, total/HDL-cholesterol ratio, total glutathione and nitric oxide.	(Asemi et al. 2016)

3.2. Lycopene

Lycopene is a carotenoid pigment and a principal cause of the red color of tomato and watermelon, which is used as an important indicator of tomato quality. As mentioned, lycopene has no vitamin A activity (Rao, Ray, and Rao 2006; van Breemen and Pajkovic 2008). By absorbing light during photosynthesis, this pigment protects plants against photosensitization (Mazza 1998). In addition to tomatoes and tomato products, watermelon, papaya and pink grapefruit are considered other important sources of dietary lycopene (Gross 1991; Mangels et al. 1993).

Since lycopene is fat-soluble, it is primarily concentrated in fatty tissues like adrenal glands, liver and testis (Khachik et al. 2002; Rao and Agarwal 2000). Possibly, the major physiological effects of lycopene in these tissues are antioxidant activity through ROS inactivation and free radicals quenching (Britton 1995; Boileau et al. 1999). The antioxidant effects of lycopene are not limited to O_2^-, but can also be against lipid peroxidation and the highly destructive hydroxyl radical $HO^·$; henceforth it can be advantageous in preventing and restricting the risk of many diseases (Sies, Stahl, and Sundquist 1992). Other recognized *in vitro* effects of lycopene include reducing the oxidative damage of DNA, proteins, lipids and other cells (Di Mascio, Kaiser, and Sies 1989).

4. Phytoestrogens

Phytoestrogens are polyphenol compounds with both estrogenic and anti-estrogenic activity. Isoflavones, coumestans and lignans belong to this group of phytochemicals found in various fruits and vegetables. Soybean is one of the most important dietary sources of phytoestrogens. Flaxseed is another important source of phytoestrogens due to the high content of lignans.

4.1. Lignans

Lignans are other groups of phytoestrogens, a subgroup of polyphenols, that flaxseed is the richest dietary source, although they are also found in the bran or seed coat of cereal grains (Paliyath, Bakovic, and Shetty 2011; Thompson et al. 1991).

Enterodiol and enterolactone are two major mammalian lignans that are produced from plant lignan precursors by the action of gut microflora (Setchell 1995; Setchell et al. 1981). Except for the gastrointestinal tract, free lignans can also be released upon the hydrolysis of lignan complex with acid or alkali. Lignans can act as a precursor of mammalian lignans, except they are potent antioxidants. Secoisolariciresinol diglucoside (SDG) in flaxseed and matairesinol are two important lignans whose beneficial effects are well established in human health (Paliyath, Bakovic, and Shetty 2011).

5. Phytosterols and Phytostanols

Since plant sterols serve as cell wall structural components, roughly speaking, they can be found in nearly all plant materials. Stanols are a saturated form of sterols, with no carbon-carbon double bond, while cholesterol and most natural phytosterols are considered unsaturated since they contain one or two carbon-carbon double bonds (Moreau 2003; Hicks and Moreau 2001; Miettinen and Gylling 2004; Piironen et al. 2000). The role of sterols is similar to that of cholesterol in animals, which regulates cell membranes' fluidity and permeability in vegetable oil, seeds, and nuts (Verleyen et al. 2002).

The biochemical structure of phytosterols is similar to that of cholesterol. Hence, intake of phytosterol and phytostanol can reduce serum cholesterol levels. The proposed mechanisms for this effect include; (1) displacing cholesterol during micelles formation and consequently reducing the intestinal absorption of cholesterol, (2) limiting cholesterol solubility in the intestine due to their higher hydrophobicity and (3) retaining cholesterol in the esterified form, due to inhibiting the intestinal cholesterol ester hydrolysis (Ling and Jones 1995).

Sitosterol, campesterol and stigmasterol are considered as the most abundant phytosterols, respectively (Ellegård et al. 2007; Moreau, Whitaker, and Hicks 2002), which are mainly obtained through dietary vegetable oils (Weihrauch and Gardner 1978; Piironen et al. 2000), which may contain 0.1–1.0% of sterols (Verleyen et al. 2002). Sitosterol is the most widely used plant sterol in the diet, accounting for about 56–79% of the total sterol content, followed by campesterol (20%) and stigmasterol (10%) (Normén et al. 2001; Morton et al. 1995; de Vries et al. 1997). The amount of dietary stanols is estimated to be 9% (Normén et al. 2001). Cereals, nuts, vegetables, berries and other fruits also contain small amounts of sterols, which are reported to be in the range of 0.05–0.22 g/100 g (Piironen et al. 2000). At present, free phytosterols extracted from various plant sources are extensively used in fortified foods and dietary supplements. Some food products enriched with different sterols and stanols are shown in Table 3.3.

Table 3.3. Selected publications on incorporation of phytosterols into food products

Food product	Type of phytosterol	Major findings	Reference
Milk-based fruit beverage	β-sitosterol and campesterol	The randomized, double-blind, crossover study (performed in 36 postmenopausal women) showed that the β-sitosterol and campesterol serum concentrations increased after four weeks of consuming the beverages (1.5 g sterol-enriched product/day), thereby confirming suitability as matrices for sterol enrichment.	(Garcia-Llatas et al. 2015)
Yogurt	β-sitosterol, stigmasterol, campsterol and brassicasterol	β-sitosterol, stigmasterol, campsterol and brassicasterol incorporation of phytosterols into yogurt led to lower viscosity, syneresis, and higher firmness than control samples. Sensory results showed no significant difference between enriched and control samples regarding texture, appearance, flavor and overall acceptance.	(Izadi et al. 2015)
Instant black tea	β-sitosterol, sitastanol, campesterol, campestenol, and stigmasterol	Consumption of black tea enriched with phytosterols (2 g phytosterols/d) for 4 weeks decreased total cholesterol, low-density lipoprotein (LDL) cholesterol, apolipoprotein B, and oxidative stress index, while increasing the concentrations of adiponectin, total antioxidant status, and tissue-plasminogen activator.	(Orem et al. 2017)
Margarine	β-sitosterol esters	Margarine with unsaturated ester oxidized faster than those with saturated counterpart and native β-sitosterol, particularly at higher temperature (55°C).	(Panpipat, Chaijan, and Guo 2018)
Butter	β-sitosterol-loaded nanostructured lipid carriers	Incorporation of encapsulated β-sitosterol into butter, showed good stability during three months' storage period. No change was observed in both acid and peroxide values and the antioxidant property.	(Bagherpour et al. 2017)
Deli-style turkey	β-sitosterol, campesterol, stigmasterol and brassicasterol	Plant sterols remained stable during the seven-day storage period. Sterols affected the texture parameters, shear force, lipid oxidation, L values, and water-holding capacity. Results of sensory analysis indicated a slight difference between control and enriched samples.	(Grasso et al. 2016)

6. Probiotics, Prebiotics and Synbiotics

6.1. Probiotics

Probiotics are defined as "live microorganisms which when administered in adequate amounts confer a health benefit to the host" (Kerry et al. 2018). Several species of the genera *Lactobacillus*, *Bifidobacterium*, *Streptococcus*, *Lactococcus*, *Bacillus*, *Streptococcus*, *Pediococcus*, *Enterococcus*, *Bacteroides*, *Akkermansia*, *Propionibacterium* and *Saccharomyces* are recognized as probiotics (Khorshidian et al. 2020) (Table 3.4). Although there is no consensus on the minimum of viable probiotic cells per gram or milliliter of probiotic products, generally, the concentrations of 10^6 and 10^7–10^8 CFU/mL or CFU/g have been accepted as the minimum and satisfactory levels, respectively. Furthermore, it has been pointed out that probiotic products should be consumed regularly with an approximate amount of 100 g/day to deliver about 10^9 viable cells into the intestine (Mohammadi, Sohrabvandi, and Mortazavian 2012). Probiotics exhibit several health benefits, including antimicrobial, antidiabetic, anti-obesity, anti-inflammatory, anticancer, antiallergic, and angiogenic activities (Kerry et al. 2018).

Recently, it has been suggested a different definition of probiotic and mentioned that probiotic refers to viable or non-viable probiotic cells (vegetative or spore; intact or ruptured) that are potentially healthful to the host. Based on the latter definition, probiotics are categorized into three groups; including 'true probiotic' (TP: viable and active probiotic cells), pseudoprobiotic (PP: viable but inactive viable probiotic cells), and ghost probiotic' (GP_I or GP_R: intact dead probiotic cell or ruptured dead probiotic cell). True probiotic should tolerate the environmental conditions and keep its activity in the contributed products (food, drug, or supplement) and/or in the biological systems (i.e., human gastrointestinal tract). Based on the site of action of probiotics, these scientists categorized each class of probiotic (TP, PP, GP_I, and GP_R) into two groups, including internal (*in vivo*: I) and external (*in vitro*: E). Based on this novel conceptualization, probiotics can be also categorized as 'viable probiotic' (VP) in the forms of active (VPI) or inactive (IVP), or 'nonviable probiotic' (NVP) in the forms of intact cell (NVPI) or ruptured cell (NVPR). All mentioned forms provide their specific health promoting effects (Zendeboodi et al. 2020; Zendeboodi and Mortazavian 2021).

In many functional foods and dietary supplements, bacterial species with probiotic properties are widely used, particularly *Bifidobacterium* and *Lactobacillus* strains (Plaza-Diaz et al. 2014; Heydari et al. 2018). Although probiotics confer numerous health benefits on the host when administered in sufficient quantity, their most important effect is the improvement of gastrointestinal health (Ray and Didier 2014) (Table 3.5). Not only is the gut health enormously influenced by intestinal microbiota, but gut microbial colonization also affects immune function and overall body health (Cummings et al. 2004). The gut microbiota is associated with energy regulation because of its ability to alter nutrient absorption.

Following the effects of microbial populations on intestinal permeability and secretion of gut hormones and gastrointestinal function and motility, digestion, absorption and metabolism of nutrients, compounds and drugs are changed. Another important function of intestinal microbiota is reinforcing the intestinal barrier against opportunistic and pathogenic microorganisms. In addition, synthesis of B group and K vitamins, the production of short-chain fatty acids (SCFAs) and the regulation of pro- or anti-inflammatory state are other activities of the intestinal microflora (Cummings et al. 2004; Gibson and Roberfroid 1995; Watson and Preedy 2012).

Based on mentioned activities, the positive and therapeutic effects of probiotics in antibiotic- associated diarrhea (McFarland 2007; Chmielewska and Szajewska 2010; Hempel et al. 2012; Russo, Linsalata, and Orlando 2014; Isolauri, Rautava, and Salminen 2012), constipation (McFarland 2007; Chmielewska and Szajewska 2010; Hempel et al. 2012; Russo, Linsalata, and Orlando 2014; Isolauri, Rautava, and Salminen 2012), lactose intolerance, gastroenteritis, irritable bowel syndrome (IBS) and inflammatory bowel disease (IBD) have been revealed (Marteau and Shanahan 2003; Brown and Valiere 2004; Stanton et al. 2003). As discussed earlier, the effects of probiotics are not limited to the gastrointestinal tract, but the benefits of taking probiotics in problems such as food allergies (McFarland 2007; Chmielewska and Szajewska 2010; Hempel et al. 2012; F. Russo, Linsalata, and Orlando 2014; Isolauri, Rautava, and Salminen 2012), cancer (McFarland 2007; Chmielewska and Szajewska 2010; Hempel et al. 2012; F. Russo, Linsalata, and Orlando 2014; Isolauri, Rautava, and Salminen 2012), genito-urinary tract infections and immune deficiency (Marteau and Shanahan 2003; Brown and Valiere 2004; Stanton et al. 2003) are also evident.

The role of probiotics in reducing inflammation has been well documented. Gut microbiota modification and gut-associated lymphoid immune system regulation are the two major components of this mechanism.

Probiotics can suppress inflammation by modulating important signaling pathways such as toll-like receptors and nuclear factor-κB (Hotel and Cordoba 2001; Hill et al. 2014; Bermudez-Brito et al. 2012; Fontana et al. 2013).

Table 3.4. The most common microorganism used as probiotics (Zendeboodi et al. 2020)

Probiotic genus	Species
Lactobacillus	*L. plantarum, L. paracasei, L. acidophilus, L. casei, L. rhamnosus, L. crispatus, L. gasseri, L. reuteri, L. bulgaricus*
Bifidobacterium	*B. adolescentis, B. animalis, B. bifidum, B. infantis, B. lactis, B. longum, B. catenulatum*
Propionibacterium	*P. jensenii, P. freudenreichii*
Pediococcus	*P. acidilactici, P. pentosaceus*
Leuconostoc	*Lc lactis subsp. cremoris, Lc. mesenteroides*
Enterococcus	*E. faecium, E. faecalis*
Streptococcus	*S. oralis, S. uberis, S. rattus, S. salivarius, S. sanguis, S. mitis, S. thermophilus*
Bacillus	*B. coagulans, B. subtilis, B. laterosporus*
Bacteroides	*B. uniformis*
Akkermansia	*A. muciniphila*
Saccharomyces	*S. cerevisiae, S. boulardii*

6.2. Prebiotics

Prebiotics are defined as non-digestible food components that offer beneficial effects, especially on the gastrointestinal system, by selectively modulation of the gut microbiota favorable to the host (Roberfroid 2000; Delzenne, Neyrinck, and Cani 2013; Parnell et al. 2012; Pineiro et al. 2008). Prebiotics are mostly derivatives of carbohydrates that, because they are not digested in the upper digestive system, are delivered virtually intact to the colon, where they are fermented by microflora (Ray and Didier 2014). Figure 3.5 demonstrates different types of prebiotics.

Prebiotics can be dietary fiber, but not all fibers are necessarily prebiotic. Asparagus, garlic, leeks and onions are rich sources of prebiotics (Roberfroid 2002). Inulin, oligofructose, and fructooligosaccharide (FOS) appeared to have the most bifidogenic effect. Prebiotics selectively stimulate the growth and/or activity of protective bacteria mainly bifidobacteria and lactobacilli and also cause natural resistance to invasive pathogens; which is why consumption of these foods is associated with an increase in fecal lactobacilli and also decrease pathogens such as clostridia species (Meyer 2015; Shah 2004).

The average daily consumption of prebiotic fiber is estimated to be 1-4 g in the United States and 3-11 g in Europe (Roberfroid 2007b). Currently, commercially available prebiotics includes lactulose, lactitol, fructooligosaccharides (FOS) and galacto-oligosaccharides (GOS) (Malaguarnera et al. 2014).

Several criteria have been proposed for prebiotics (Laparra and Sanz 2010; Manning and Gibson 2004; Marcel Roberfroid 2007a; Gibson and Roberfroid 1995; Roberfroid 2001), but what most researchers agree on include resistance to low acidity, digestion and absorption, susceptible to fermentation by gastrointestinal microflora, exposing stimulatory effects on targeted growth of species of favorable intestinal bacteria, production of SCFAs and reducing the pH of intestinal contents as well as being safe for human health. The two carbohydrates that fulfill these criteria are inulin-type fructans and GOS, although investigation on other prebiotics continues (Roberfroid et al. 2010).

Prebiotics can be incorporated into foods to increase the microflora growth in the GI tract and the quality of food products (Hosseinvand et al. 2019). Several prebiotic food products, including dairy products, beverages, bakery products, snacks and meat products, have been formulated that are summarized in Table 3.6.

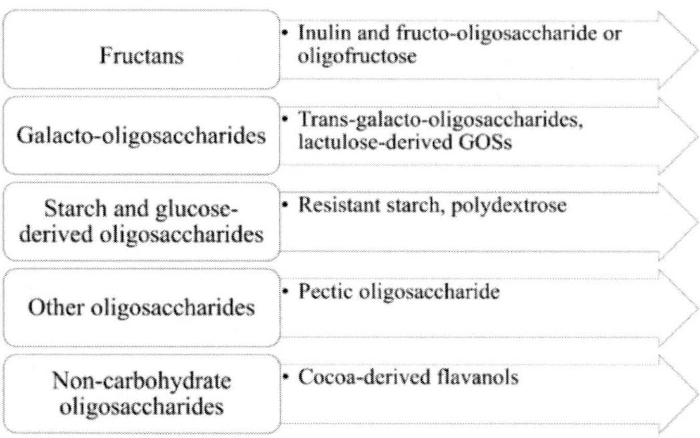

Figure 3.5. Different types of prebiotics.

Table 3.5. Selected clinical experiments regarding health effects of probiotics

Probiotic strain	Subjects	Time of administration	Main findings	Reference
B. breve CBT BR3 *L. plantarum* CBT LP3	50 subjects with body mass index over 25 kg/m^2	12 weeks	Administration of probiotics improved obesity-related markers in obese people.	(Song et al. 2020)
L.acidophilus CUL60, *L. acidophilus* CUL21, *L. plantarum* CUL66, *B. bifidum* CUL20 *B. animalis* subsp. *lactis*	220 Bulgarian participants (30 to 65 years old) with BMI 25-34.9 kg/m^2	6 months	Consumption of probiotics reduced bodyweight, BMI, waist circumference and waist-to-height ratio in a free-living overweight/obese population with greater weight loss observed in the overweight and the female participants.	(Michael et al. 2020)
B. lactis UBBLa-70	32 adult women with BMI ranging from 30 to 34.9 kg/m^2	8 weeks	Changes in metabolites related to improved metabolic health was observed.	(Crovesy, El-Bacha, and Rosado 2021)
L. acidophilus, L. rhamnosus, *L. paracasei, P. pentosaceus,* *B. lactis, B. breve*	38 obese nonalcoholic fatty liver disease patients	12 weeks	Administration of probiotics was effective in reducing triglycerides as well as intrahepatic fat.	(Ahn et al. 2019)
L. paracasei DSM 24733, *L. plantarum* DSM 24730, *L. acidophilus* DSM 24735, *L. delbrueckii* subsp. *bulgaricus* DSM 24734, *B. longum* DSM 24736, *B. infantis* DSM 24737, *B. breve* DSM 24732, *S. thermophilus* DSM 24731	39 patients with NAFLD	12 months	A significant improvement in liver histology, ALT and cytokines was observed.	(Duseja et al. 2019)

Probiotic strain	Subjects	Time of administration	Main findings	Reference
L. acidophilus, *L. casei* *B. bifidum*	60 subjects with gestational diabetes	6 weeks	Probiotic supplementation had beneficial effects on fasting plasma glucose, serum high-sensitivity C-reactive protein, plasma total antioxidant capacity, malondialdehyde and oxidative stress index, but did not affect pregnancy outcomes.	(Badehnoosh et al. 2018)
L. rhamnosus *B. animalis* subsp. *lactis*	411 overweight and obese women	28 weeks	Probiotics did not prevent gestational diabetes mellitus in overweight and obese pregnant women.	(Callaway et al. 2019)
L. paracasei *L. salivarius* *L. plantarum*	50 Vietnamese patients with unconstipated IBS	4 weeks	The mixture of Lactobacilli was effective in the global relief of IBS symptoms, as well as in relieving abdominal pain without significant adverse events.	(Oh et al. 2019)
B. longum BB536 *L. rhamnosus* HN001	23 lactose intolerant patients	30 days	Probiotic treatment alleviated symptoms in subjects with lactose intolerance and persistent functional gastrointestinal diseases through a positive modulation of gut microbial composition and relative metabolism.	(Vitellio et al. 2019)
L. acidophilus *L. plantarum*	25 subjects suffering from colorectal cancer, 28 with polyps and 24 healthy subjects	-	The results indicated that taking *Lactobacillus acidophilus* may be a way of preventing, treating or reducing the severity of CRC.	(Zinatizadeh et al. 2018)

6.3. Synbiotics

The term synbiotics has been adopted to refer to a combination of prebiotics and probiotics, which provides a synergistic benefit (Pineiro et al. 2008). Undoubtedly taking probiotics and prebiotics simultaneously will have additive health benefits. Combining both types of components has eventually resulted in the introduction of so-called synbiotic products (Roberfroid 2000).

The first aspect that should be considered during the development of synbiotic formula is the selection of suitable probiotics and prebiotics that exhibit positive effects on the host's health when used individually. A prebiotic should selectively stimulate the growth of microorganisms, having a beneficial effect on health without promoting the growth of other microorganisms (Zakipour et al. 2020). It has been indicated that using prebiotics resulted in increased resistance of probiotics to environmental conditions such as oxygenation, pH, and temperature in the intestine of a particular organism (Sekhon and Jairath 2010). The probiotic strains used in synbiotic formulations include lacbobacilli, bifidobacteria spp., *Saccharomyces boulardii*, *Bacillus coagulans* etc., while the major prebiotics used include oligosaccharides like fructooligosaccharide, galactooligo-saccharides and xyloseoligosaccharide, inulin, prebiotics from natural sources like chicory and yacon roots, etc. The health benefits attributed to synbiotics consumption by humans include: 1) Increase in population of lactobacilli and bifidobacteria and promoting balanced gut microbiota, 2) Improvement of liver function in cirrhotic patients, 3) Improvement of immunomodulating ability, 4) Prevention of bacterial translocation and reduced incidences of nosocomial infections in surgical patients and similar interventions (Zhang et al. 2010). Several studies and clinical trials have been conducted to investigate the effect of synbiotics on the improvement of human health and treatment of metabolic syndrome, IBS, IBD, colon cancer, diarrhea, kidney and liver diseases (Krumbeck, Walter, and Hutkins 2018).

7. Macronutrients

7.1. Bioactive Carbohydrates

Bioactive carbohydrates include mainly dietary fiber, natural polysaccharides and prebiotics (Hayes, 2013; Khorshidian et al. 2018). These compounds can be extracted from natural sources such as lower and higher plants, animals and microorganisms (Figure 3.6). Researchers have declared various biotechnological and pharmaceutical applications for these compounds. For example, chitin, a polysaccharide extracted primarily from prawn and crab shell material, and its de-acetylated form called chitosan, are considered functional foods as they can have anti-obesity and satiety effects (Hayes et al. 2008). Another general application is in encapsulation technologies (Hayes et al. 2008; Yousefi et al. 2019) and manufacturing medical bandages (Hayes et al. 2008). More recently, the prebiotic effects of chitin and chitosan have been studied (Lordan, Ross, and Stanton 2011; O'Sullivan et al. 2010).

Dietary fiber is defined as carbohydrate resistant to digestion and absorption, partially fermented by gastrointestinal microbiota. In addition, dietary fibers are considered as bioactive carbohydrates that can increase the number of beneficial gastrointestinal bacteria.

The positive effects of dietary fiber on various aspects of health, especially in reducing the risk of certain chronic diseases, including gastrointestinal disorders (Petruzziello et al. 2006), obesity (Lairon et al. 2005), type 2 diabetes (Montonen et al. 2003) and cardiovascular disease (Liu et al. 1999) have been proved (Watson and Preedy 2010; Wildman 2016).

Although the recommendations for dietary fiber intake are different depending on age, sex and total energy intake, the general recommendation for a daily adequate intake (AI) is 14 g/1000 kcal (McGuire 2011).

Given the low dietary fiber intake in many people, fortifying foods with fiber can be an effective strategy to promote community health, which requires more investigation. In the same vein, polysaccharides derived from plants, cereals and soybeans can be added to various foods, including dairy products (Dhingra et al. 2012). The characteristics and functions of fibers differ according to their chemical, physical, and functional properties. Soluble fibers like pectin, gums, inulin-type fructans, which are fermented in the colon, can influence the turnover of the gut microflora. Fruits and vegetables are rich sources of soluble fibers. Soluble fibers are also called fermentable fiber or viscose fiber. On the other hand, insoluble fibers like wheat bran have bulking action and reduce the transit time. They consist of lignin, cellulose and some hemicelluloses. Insoluble fibers are found mainly in cereals, nuts, and seeds (Anderson et al. 2009; Lattimer and Haub 2010; de Mattos and Martins 2000).

It has been expressed that polysaccharides exhibit antioxidant activity dependent on their structure. The major influencing physicochemical properties on antioxidant effects include water-solubility, molecular weight, presence of backbone chain of β-(1→3)-D-glucose and β-(1→6)-side chain, degree of branching, functional groups and triple-helical structure (Yuan et al.

2015). Antioxidant activity of polysaccharides extracted from *Allium sativum* (Cheng and Huang 2018), ginger (Yun Wang et al. 2018), ginseng (F. Chen and Huang 2019), *Chlorella vulgaris* (Yu et al. 2019), *Porphyra haitanensis* (Khan et al. 2020) and *Desmodesmus armatus* (Cao et al. 2020) have been reported.

Table 3.6. Selected publications on incorporation of prebiotics into food products

Food product	Prebiotic type	Main effects	Reference
Fermented sausage	β-glucan (0, 1, 2%)	Addition of β-glucan decreased thiobarbituric acid reactive substances values, water activity and pH values compared to control groups. The growth of lactic acid bacteria increased during fermentation and storage periods.	(Yuca et al. 2019)
strawberry-flavored whey beverage	Xyloligoosaccharide (1.25%)	Addition of xyloligoosaccharide resulted in a beverage with better functional characteristics, rheological properties (higher viscosity), volatiles profile, and sensory acceptance besides providing adequate concentrations of the prebiotic in the product.	(Souza et al. 2019)
Bread	β-glucan (0.8, 1, 1.2% w/w) and resistant starch (5.5, 8, 10.5% w/w) and combination of two prebiotics (0.5 β-glucan and 4% resistant starch)	β-glucan and resistant starch affected dough rheological properties significantly while not affecting the sensory properties of bread. Both of them were found to increase the shelf life of bread and decrease its staleness.	(Mohebbi et al. 2018)
Gluten-free biscuit	Inulin and resistant dextrin (25, 50 and 75 g/100 g flour basis)	There were significant differences in acceptability between the control and the biscuits with 25% of fat replaced by resistant dextrin and inulin except for color and flavor that were better than the control. Gluten-free biscuits containing 25 g inulin and 25 g resistant dextrin were similar to the control biscuits.	(Emami et al. 2018)
Yogurt	Inulin, lactulose, lactitol, Hi-maize, maltodextrin and β-glucan (1.5 or 3%)	β-glucan-1.5 and Hi-maize-1.5 can be used successfully as prebiotic compounds in probiotic yogurt because they provided better rheological, syneresis, and sensory properties.	(Heydari et al. 2018)

Food product	Prebiotic type	Main effects	Reference
sheep milk ice cream	Inulin, fructo-oligosaccharide, galactooligosaccharide, short-chain fructo-oligosaccharide, resistant starch, soluble corn fiber, and polydextrose (10% w/w)	Addition of the prebiotic inulin and fructo-oligosaccharide have proven to be a promising alternative as fat substitutes in sheep milk ice cream formulation, due to the similar rheological properties and sensory attributes.	(Balthazar et al. 2017)
Sausage	Resistant starch (4%)	The addition of resistant starch as a fat substitute in sausages did not influence the composition, texture profile analysis, and sensory analysis, indicating it to be a promising ingredient in making healthier meat products.	(Garcia-santos et al. 2019)
Soursop flavored whey beverage	Inulin (6%)	Inulin had a great effect on physical stability. Still without a stabilizer, a higher degree of polymerization (DP) was more effective and the influence of DP reduced when gellan gum was used.	(Guimarães et al. 2018)
Full-fat yogurt	β-glucan (0.5-2% w/v)	Addition of β-glucan significantly improved whey separation (syneresis), viscosity, texture profile and sensory characteristics during storage.	(Kaur and Riar 2020)
Greek yogurt	Xylooligosaccharide, galactooligosaccharide, polydextrose, fructooligosaccharide or inulin (2.5%)	Galactooligosaccharide, polydextrose, and inulin proved to be an effective technological alternative for the development of Greek yogurts.	(Costa et al. 2019)

Another promising feature of polysaccharides is an immunomodulatory activity associated with their structures. It has been elucidated that polysaccharides with a configuration of β-1-3, 1-4, or 1-6 branch chains are essential for activity while complex branch-chained polysaccharides with anionic structures and higher molecular weights have greater immunostimulating activities (Kim and Kim 2017). It has been stated that plant polysaccharides interact with specific receptors on cells, stimulate macrophages and enhance the immune system efficiency (Ma et al. 2015). Bioactive carbohydrates exhibit antitumor activity by directly affecting tumor cells and enhancing the immune function of the organism (Pandya, Dhuldhaj, and Sahay 2019).

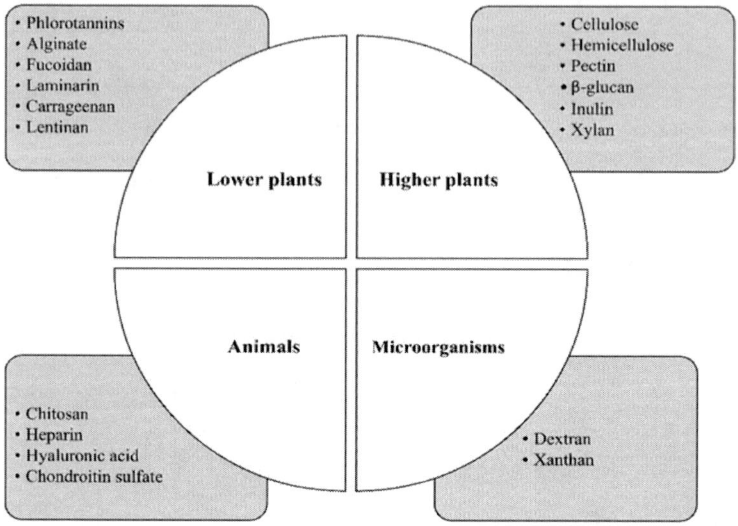

Figure 3.6. Different sources of bioactive carbohydrates.

Several plant polysaccharides, including cellulose, protein-bound polysaccharides, glucomannan, mannose guar gum, pectin/pectin fibers, caryophyllene, and mucilaginous fiber, have been reported to have an anti-diabetic effect (Wang et al. 2013). These polysaccharides can initiate insulin secretion, carbohydrate digestion and absorption (Liu, Willför, and Xu 2015).

7.2. Bioactive Lipids

Several bioactive lipid components such as fat-soluble vitamins, phytosterols, carotenoids, and polyunsaturated fatty acids have been advantageous for human health. It has been recommended that dietary lipids should provide around 25-35% of the total energy requirements, of which up to 20% consists of monounsaturated fatty acids (MUFAs), 10% PUFAs (polyunsaturated fatty acids) and less than 10% saturated fatty acids (SFAs). It seems many systematic effects of lipids are due to the eicosanoid pathway (Trumbo et al. 2002; Lichtenstein et al. 2006; Kris-Etherton, Innis, and Ammerican 2007; Gebauer et al. 2006; Kris-Etherton et al. 2000; Zello 2006). Functional and bioactive lipids may be extracted from plants, animals and microorganisms. A major part of the nutraceutical effects of fatty acids are related to the eicosanoids and prostaglandins synthesis pathway. Prostaglandins are a diverse group of hormones and mediators regulating physiological functions. For example, it has been shown that prostaglandins in the gastrointestinal tract play a protective role in the gastric mucosa and a regulatory role in gastrointestinal motility. In addition, the effect of prostaglandins in inflammatory bowel disease and colorectal cancer has been reported (Milner et al. 2010; Ahrenstedt and Knutson 1994; Eberhart et al. 1994). Among fatty acids, beneficial health effects of medium-chain fatty acids (MCFAs), omega-3 PUFAs, and conjugated fatty acids are mainly studied (Shahidi 2010). Medium-chain fatty acids are a subgroup of saturated fatty acids with 6–12 carbon atoms. MCFAs are of nutritional interest and are used in malabsorption, maldigestion, metabolic disorders allied to cystic fibrosis, inflammatory bowel disease and enteritis (Pfeuffer and Schrezenmeir 2002; Kabara 2000). Conjugated linoleic acid (CLA) is an isomeric group of linoleic acid containing conjugated dienes. CLA is a trivial part of ruminant animal fats and is mainly present in meat and dairy products. Cheeses are the richest source of CLA. The most abundant natural isomer is 9-cis, 11-trans isomer (Shahidi 2010; Whigham, Watras, and Schoeller 2007). Table 3.7 presents major bioactive lipids and their health benefits.

Table 3.7. Major dietary bioactive lipids and their health benefits

Bioactive lipid	Dietary source	Health effects
Fatty acids (butyric acid, medium chain fatty acids, long chain fatty acids, CLA, arachidonic acid	Milk, egg, fish, vegetable oils, tallow	Prevention and treatment of obesity, dementia and Alzheimer's disease, controlling colorectal cancer, lowering blood glucose and cholesterol antihypertensive effects and decreasing the risk of adverse cardiovascular
Carotenoids	Fruits and vegetables, vegetable oils	Inhibition of atherosclerosis, age-related muscular degeneration, and multiple sclerosis, coronary heart disease and breast cancer
Phytosterols	Vegetable oils, algae	Reducing total and LDL cholesterol, anti-carcinogenic, anti-inflammatory, anti-oxidative and immunomodulatory effects, prevention of cardiovascular disease
Fat-soluble vitamins (vitamins A, D, E and K)	Fruits and vegetables, vegetable oils	Reduction of heart disease, delay of Alzheimer's disease and prevention of cancer

7.3. Bioactive Peptides

Proteins and peptides, apart from contributing to energy supply and the growth of the body as sources of amino acids, can also affect specific biological activities (Mine, Li-Chan, and Jiang 2011). Bioactive peptides have been defined as specific sequences of amino acids that may exert regulatory functions and positive impacts (Kitts and Weiler 2003). These biologically active peptides are produced during the enzymatic digestion in the gastrointestinal tract, or by bacterial fermentation, or food processing (Nollet and Toldrá 2012). The bioactivity of peptides is mainly dependent on the enzymes used for hydrolysis, processing conditions and the size of isolated peptides (Udenigwe and Aluko 2012). The size of the bioactive peptide sequence varies from 2 to 20 or even more amino acid residues and may possess multifunctional properties (Girgih et al. 2014).

Various proteolytic enzymes are used to hydrolyze proteins, including papain, pepsin, protease, pancreatin, trypsin, chymotrypsin and thermolysin, along with the enzymes from bacterial and fungal sources, depending upon the parent protein to produce protein hydrolysates (Bhandari et al. 2020). The biological activity and sequence of peptides vary based on the enzyme used (Mojica and De Mejía 2016). Microbial fermentation is another way of obtaining bioactive peptides achieved through starter and non-starter cultures used in industries with proteolytic nature (Girgih et al. 2014). The extent of hydrolysis depends on the type of starter culture used, protein source and fermentation time (Fitzgerald and Murray 2006). In a study carried out by Baptista et al. (2018), the effect of adding the adjunct culture *Lactobacillus helveticus* LH-B02 on the proteolysis and peptide profile of Prato cheese during ripening was investigated. The results showed that the addition of adjunct culture influenced the Prato cheese peptide profile and increased the release of peptides from β-casein and particularly the angiotensin-converting enzyme-inhibitory peptide. Corn gluten meal peptides with strong antioxidant activity have been produced by solid-state fermentation with *Bacillus subtilis* MTCC5480 (Jiang et al. 2020).

Several functionalities of dietary peptides and proteins have been recognized in various systems in the body, including the digestive system, the immune system and even the nervous system. The amino acid composition and sequence specify the activity of peptides. These peptides act is either locally in the gastrointestinal tract or after absorption and reaching the target organ. Well- known or commercially used bioactive peptides are derived from milk proteins; of course, any food protein from plant or animal origin can be a precursor of bioactive peptides (Nollet and Toldrá 2012; Lee and Ho 2002; Gobbetti, Minervini, and Rizzello 2007; Hartmann and Meisel 2007). A great number of bioactive peptides from food sources, including milk, meat, egg, maize, soybean, rice, mushroom, amaranth, pumpkin, etc. have been isolated (Sánchez and Vázquez 2017). These bioactive peptides exhibit various functionalities, including antioxidant, antimicrobial, anti-inflammatory, antihypertensive, anticancer, osteoprotective, hypolipidemic and hypocholesterolemic activities.

Bioactive peptides with Angiotensin-I converting enzyme (ACE) inhibitory activity prevent the conversion of Angiotensin-I to Angiotensin II leads to vasodilation and lowers blood pressure. In a study by Ciau-Solís, Acevedo-Fernández, and Betancur-Ancona (2018), Lima bean protein hydrolysates showed both *in vitro* and *in vivo* antihypertensive activity. In another study, fermentation of cheese whey by its native microbiota resulted in bioactive peptides with ACE-inhibitory activity (Mazorra-Manzano et al. 2020).

Antimicrobial peptides mainly contain an α-helical structures that are cationic and amphipathic and fewer numbers are hydrophobic α-helical peptides. The cationic section in peptide structure interacts electrostatically with anionic phospholipids that create pores in the cytoplasmic membrane, ionic imbalance across the membrane and eventually cell lysis (Bhandari et al. 2020). Gao et al. (2019) isolated and identified three antimicrobial peptide fractions from laba garlic which exhibited antimicrobial activity against *E. coli* and *Staphylococcus aureus.*

The effects of bioactive peptides in the gastrointestinal tract are of special importance. For example, it has been shown that peptides, such as cholecystokinin, can act as a digestive hormone to control gastrointestinal motility and secretions or contribute to appetite and satiety regulation (Foltz et al. 2008; Shimizu and Son 2007; Young and Mine 2009). In some studies, the biological activities of peptides in inflammation and oxidative stress are also noted (Mine, Li-Chan, and Jiang 2011). Several studies have discussed the health benefits of bioactive peptides summarized in Table 3.8.

Conclusion

Several bioactive compounds can be extracted from different sources, including polyphenolic compounds, carotenoids, prebiotics and probiotics, bioactive lipids, carbohydrate and peptides, which have emerged as healthful ingredients to be used in the development of functional food products and novel supplements.These substances can exert different biological activities such as antioxidant, antimicrobial, anti-inflammatory, antidiabetic and anticancer activities and prevent chronic diseases. However, further in vivo research is necessary to establish specific bioactive compounds' target site effects. Moreover, studies regarding enhancing the bioavailability and stability of bioactive compounds in food products seem indispensable.

Table 3.8. Bioactive peptides isolated from different sources and their biological activity

Biological effect	Source of bioactive peptide	Reference
Antioxidant	Buffalo yogurt	(Taha et al. 2017)
	Flathead (*Platycephalus fuscus*)	(Nurdiani et al. 2017)
	Pecan meal	(Hu et al. 2018)
	Sesame protein	(Lu et al. 2019)
	Brown rice protein	(Kunlun Liu, Du, and Chen 2020)
	Round scad protein	(Guo et al. 2020)
	Pea protein	(Ding et al. 2020)
Antihypertensive	Seaweed protein	(Admassu et al. 2018)
	Egg white protein	(Liao, Fan, and Wu 2018)
	Soybean protein	(Daliri et al. 2019)
	Salmon gelatin	(Neves et al. 2017)
	Whey protein	(Alvarado et al. 2019)
	Quinoa protein	(Guo et al. 2020)
	Peanut protein	(Gong et al. 2017)
Anticancer	Rainbow trout skin	(Yaghoubzadeh et al. 2020)
	Black soybean	(Z. Chen et al. 2019)
	Buffalo and cow milk Cheddar cheese	(Rafiq et al. 2018)
	Camel milk	(Ayyash et al. 2018)
	Amaranthus cruentus protein	(Ramkisson et al. 2020)
	Lactoferricin	(Arias et al. 2017)
Antidiabetic	Corn germ protein	(Karimi, Azizi, and Ahmadi Gavlighi 2020)
	Camel milk protein	(Mudgil et al. 2018)
	Casein	(Megrous et al. 2020)
	Walnut	(Wang et al. 2020)
	Atlantic salmon	(Harnedy et al. 2018)
	Spirulina platensis	(Hu et al. 2019)
	Potato protein	(Marthandam Asokan et al. 2019)

References

Abdul-Hamid, A., and Luan, Y. S. 2000. "Functional properties of dietary fibre prepared from defatted rice bran." *Food Chemistry* 68 (1): 15-19.

Adebooye, O. C., Alashi, A. M., and Aluko, R. E. 2018. "A brief review on emerging trends in global polyphenol research." *Journal of Food Biochemistry* 42 (4): e12519.

Admassu, H., Gasmalla, M. A. A., Yang, R., and Zhao, W. 2018. "Bioactive peptides derived from seaweed protein and their health benefits: antihypertensive, antioxidant, and antidiabetic properties." *Journal of Food Science* 83 (1): 6-16.

Afford S, R. S. 2000. "Apoptosis." *Molecular Pathology* 53 (2): 55–63.

Agarwal, S., and Rao, A. 2000. "Carotenoids and chronic diseases." *Drug Metabolism and Drug Interactions* 17 (1-4): 189-210.

Aggarwal, B. B., and Harikumar, K. B. 2009. "Potential therapeutic effects of curcumin, the anti-inflammatory agent, against neurodegenerative, cardiovascular, pulmonary, metabolic, autoimmune and neoplastic diseases." *The International Journal of Biochemistry & Cell Biology* 41 (1): 40-59.

Aggarwal, B. B., Sundaram, C., Malani, N., and Ichikawa, H. 2007. "Curcumin: the Indian solid gold." In *The Molecular Targets and Therapeutic Uses of Curcumin in Health and Disease*, 1-75. Springer.

Agrawal, Y. O., Sharma, P. K., Shrivastava, B., Ojha, S., Upadhya, H. M., Arya, D. S., and Goyal, S. N. 2014. "Hesperidin produces cardioprotective activity via PPAR-γ pathway in ischemic heart disease model in diabetic rats." *PLoS One* 9 (11): e111212.

Ahmad, I., Khalique, A., Junaid, M., Shahid, M. Q., Imran, M., and Rashid, A. A. 2020. "Effect of polyphenol from apple peel extract on the survival of probiotics in yoghurt ice cream." *International Journal of Food Science & Technology* 55 (6): 2580-2588.

Ahn, S. B., Jun, D. W., Kang, B.-K., Lim, J. H., Lim, S., and Chung, M.-J. 2019. "Randomized, double-blind, placebo-controlled study of a multispecies probiotic mixture in nonalcoholic fatty liver disease." *Scientific Reports* 9 (1): 1-9.

Ahrenstedt, Ö., and KNUTSON, L. 1994. "Jejunal release of prostaglandin E2 in Crohn's disease: Relation to disease activity and first-degree relatives." *Journal of Gastroenterology and Hepatology* 9 (6): 539-543.

Akgün, D., Gültekin-Özgüven, M., Yücetepe, A., Altin, G., Gibis, M., Weiss, J., and Özçelik, B. 2020. "Stirred-type yoghurt incorporated with sour cherry extract in chitosan-coated liposomes." *Food Hydrocolloids* 101: 105532.

Alashi, A. M., Taiwo, K. A., Oyedele, D. J., Adebooye, O. C., and Aluko, R. E. 2019. "Polyphenol composition and antioxidant properties of vegetable leaf-fortified bread." *Journal of Food Biochemistry* 43 (6): e12625.

Aluko, R. 2011. "Plant Derived Bioactives." *Comprehensive Biotechnology*: pp. 501–15.

Alvarado, Y., Muro, C., Illescas, J., Díaz, M. d. C., and Riera, F. 2019. "Encapsulation of antihypertensive peptides from whey proteins and their releasing in gastrointestinal conditions." *Biomolecules* 9 (5): 164.

Anderson, J. W., Baird, P., Davis, R. H., Jr., Ferreri, S., Knudtson, M., Koraym, A., Waters, V., and Williams, C. L. 2009. "Health benefits of dietary fiber." *Nutrition Review* 67 (4): 188-205.

Anonymous. 2002. "What are functional foods."

Antolovich, M., Prenzler, P., Robards, K., and Ryan, D. 2000. "Sample preparation in the determination of phenolic compounds in fruits." *Analyst* 125 (5): 989-1009.

Anuyahong, T., Chusak, C., and Adisakwattana, S. 2020. "Incorporation of anthocyanin-rich riceberry rice in yogurts: Effect on physicochemical properties, antioxidant activity and in vitro gastrointestinal digestion." *LWT-Food Science and Technology* 129: 109571.

Arias, M., Hilchie, A. L., Haney, E. F., Bolscher, J. G., Hyndman, M. E., Hancock, R. E., and Vogel, H. J. 2017. "Anticancer activities of bovine and human lactoferricin-derived peptides." *Biochemistry and Cell Biology* 95 (1): 91-98.

Asemi, Z., Alizadeh, S.-A., Ahmad, K., Goli, M., and Esmaillzadeh, A. 2016. "Effects of beta-carotene fortified synbiotic food on metabolic control of patients with type 2 diabetes mellitus: a double-blind randomized cross-over controlled clinical trial." *Clinical Nutrition* 35 (4): 819-825.

Awika, J. 2011. "Effect of bioactive components on dough rheology, baking and extrusion." *Fruit and Cereal Bioactives: Sources, Chemistry, and Applications,* CRC Press, New York: 337-343.

Ayyash, M., Al-Dhaheri, A. S., Al Mahadin, S., Kizhakkayil, J., and Abushelaibi, A. 2018. "In vitro investigation of anticancer, antihypertensive, antidiabetic, and antioxidant activities of camel milk fermented with camel milk probiotic: A comparative study with fermented bovine milk." *Journal of Dairy Science* 101 (2): 900-911.

Badehnoosh, B., Karamali, M., Zarrati, M., Jamilian, M., Bahmani, F., Tajabadi-Ebrahimi, M., Jafari, P., Rahmani, E., and Asemi, Z. 2018. "The effects of probiotic supplementation on biomarkers of inflammation, oxidative stress and pregnancy outcomes in gestational diabetes." *The Journal of Maternal-Fetal & Neonatal Medicine* 31 (9): 1128-1136.

Bagherpour, S., Alizadeh, A., Ghanbarzadeh, S., Mohammadi, M., and Hamishehkar, H. 2017. "Preparation and characterization of Betasitosterol-loaded nanostructured lipid carriers for butter enrichment." *Food Bioscience* 20: 51-55.

Balthazar, C., Silva, H., Cavalcanti, R., Esmerino, E., Cappato, L., Abud, Y., Moraes, J., Andrade, M., Freitas, M., and Sant'Anna, C. 2017. "Prebiotics addition in sheep milk ice cream: A rheological, microstructural and sensory study." *Journal of Functional Foods* 35: 564-573.

Baptista, D. P., Galli, B. D., Cavalheiro, F. G., Negrão, F., Eberlin, M. N., and Gigante, M. L. 2018. "Lactobacillus helveticus LH-B02 favours the release of bioactive peptide during Prato cheese ripening." *International Dairy Journal* 87: 75-83.

Barros, R. F. d., Torres, F. R., SILVA, P. H. F. d., Stringheta, P. C., Pereira, J. P. F., Paula, J. C. J. d., Cutrim, C. S., and CORTEZ, M. A. S. 2020. "Lutein as a functional ingredient in sheep milk yogurt: development, characterization and extraction recovery." *Food Science and Technology* 40: 683-690.

Bedrníček, J., Kadlec, J., Laknerová, I., Mráz, J., Samková, E., Petrášková, E., Hasoňová, L., Vácha, F., Kron, V., and Smetana, P. 2020. "Onion Peel Powder as an Antioxidant-Rich Material for Sausages Prepared from Mechanically Separated Fish Meat." *Antioxidants* 9 (10): 974.

Bermudez-Brito, M., Plaza-Díaz, J., Muñoz-Quezada, S., Gómez-Llorente, C., and Gil, A. 2012. "Probiotic mechanisms of action." *Annals of Nutrition and Metabolism* 61 (2): 160-174.

Bhandari, D., Rafiq, S., Gat, Y., Gat, P., Waghmare, R., and Kumar, V. 2020. "A review on bioactive peptides: Physiological functions, bioavailability and safety." *International Journal of Peptide Research and Therapeutics* 26 (1): 139-150.

Blumberg, J. B., Bolling, B. W., Chen, C., and Xiao, H. 2015. "Review and Perspective on the Composition and Safety of Green Tea Extracts." *European Journal of Nutrition & Food Safety* 5(1): 1-31.

Boileau, A. C., Merchen, N. R., Wasson, K., Atkinson, C. A., and Erdman, J. W. 1999. "Cis-lycopene is more bioavailable than trans-lycopene in vitro and in vivo in lymph-cannulated ferrets." *The Journal of Nutrition* 129 (6): 1176-1181.

Boots, A. W., Haenen, G. R., and Bast, A. 2008. "Health effects of quercetin: from antioxidant to nutraceutical." *European Journal of Pharmacology* 585 (2): 325-337.

Bora, P., Ragaee, S., and Abdel-Aal, E.-S. M. 2019. "Effect of incorporation of goji berry by-product on biochemical, physical and sensory properties of selected bakery products." *LWT* 112: 108225.

Britton, G. 1995. "Structure and properties of carotenoids in relation to function." *The FASEB Journal* 9 (15): 1551-1558.

Brown, A. C., and Valiere, A. 2004. "Probiotics and medical nutrition therapy." *Nutrition in Clinical Care: An Official Publication of Tufts University* 7 (2): 56.

Callaway, L. K., McIntyre, H. D., Barrett, H. L., Foxcroft, K., Tremellen, A., Lingwood, B. E., Tobin, J. M., Wilkinson, S., Kothari, A., and Morrison, M. 2019. "Probiotics for the prevention of gestational diabetes mellitus in overweight and obese women: findings from the SPRING double-blind randomized controlled trial." *Diabetes Care* 42 (3): 364-371.

Canalis, M. B., Baroni, M., León, A., and Ribotta, P. 2020. "Effect of peach puree incorportion on cookie quality and on simulated digestion of polyphenols and antioxidant properties." *Food Chemistry* 333: 127464.

Cao, G., Muccitelli, H. U., Sánchez-Moreno, C., and Prior, R. L. 2001. "Anthocyanins are absorbed in glycated forms in elderly women: a pharmacokinetic study." *The American Journal of Clinical Nutrition* 73 (5): 920-926.

Cao, M., Wang, S., Gao, Y., Pan, X., Wang, X., Deng, R., and Liu, P. 2020. "Study on physicochemical properties and antioxidant activity of polysaccharides from Desmodesmus armatus." *Journal of Food Biochemistry* 44 (7): e13243.

Carlo, G., Izzo, A., Maiolino, P., Mascolo, N., Viola, P., Diurno, M., and Capasso, F. 1993. "Inhibition of Intestinal Motility and Secretion by Flavonoids in Mice and Rats: Structure-activity Relationships." *Journal of Pharmacy and Pharmacology* 45 (12): 1054-1059.

Cha, J. W., Piao, M. J., Kim, K. C., Yao, C. W., Zheng, J., Kim, S. M., Hyun, C. L., Ahn, Y. S., and Hyun, J. W. 2014. "The polyphenol chlorogenic acid attenuates UVB-mediated oxidative stress in human HaCaT keratinocytes." *Biomolecules & Therapeutics* 22 (2): 136.

Chanet, A., Milenkovic, D., Manach, C., Mazur, A., and Morand, C. 2012. "Citrus flavanones: what is their role in cardiovascular protection?" *Journal of Agricultural and Food Chemistry* 60 (36): 8809-8822.

Chen, F., and Huang, G. 2019. "Antioxidant activity of polysaccharides from different sources of ginseng." *International Journal of Biological Macromolecules* 125: 906-908.

Chen, X., Yang, X., Liu, T., Guan, M., Feng, X., Dong, W., Chu, X., Liu, J., Tian, X., and Ci, X. 2012. "Kaempferol regulates MAPKs and NF-κB signaling pathways to attenuate LPS-induced acute lung injury in mice." *International Immunopharmacology* 14 (2): 209-216.

Chen, Z., Li, W., Santhanam, R. K., Wang, C., Gao, X., Chen, Y., Wang, C., Xu, L., and Chen, H. 2019. "Bioactive peptide with antioxidant and anticancer activities from black soybean [Glycine max (L.) Merr.] byproduct: isolation, identification and molecular docking study." *European Food Research and Technology* 245 (3): 677-689.

Chen, Z.-Y., Zhu, Q. Y., Tsang, D., and Huang, Y. 2001. "Degradation of green tea catechins in tea drinks." *Journal of Agricultural and Food Chemistry* 49 (1): 477-482.

Cheng, H., and Huang, G. 2018. "Extraction, characterisation and antioxidant activity of Allium sativum polysaccharide." *International Journal of Biological Macromolecules* 114: 415-419.

Chmielewska, A., and Szajewska, H. 2010. "Systematic review of randomised controlled trials: probiotics for functional constipation." *World Journal of Gastroenterology: WJG* 16 (1): 69.

Choudhury, M., Badwaik, L. S., Borah, P. K., Sit, N., and Deka, S. C. 2015. "Influence of bamboo shoot powder fortification on physico-chemical, textural and organoleptic characteristics of biscuits." *Journal of Food Science and Technology* 52 (10): 6742-6748.

Chung, K.-T., Wei, C.-I., and Johnson, M. G. 1998. "Are tannins a double-edged sword in biology and health?" *Trends in Food Science & Technology* 9 (4): 168-175.

Chung, K.-T., Wong, T. Y., Wei, C.-I., Huang, Y.-W., and Lin, Y. 1998. "Tannins and human health: a review." *Critical Reviews in Food Science and Nutrition* 38 (6): 421-464.

Ciau-Solís, N. A., Acevedo-Fernández, J. J., and Betancur - Ancona, D. 2018. "In vitro renin–angiotensin system inhibition and in vivo antihypertensive activity of peptide fractions from lima bean (Phaseolus lunatus L.)." *Journal of the Science of Food and Agriculture* 98 (2): 781-786.

Conquer, J., Maiani, G., Azzini, E., Raguzzini, A., and Holub, B. 1998. "Supplementation with quercetin markedly increases plasma quercetin concentration without effect on selected risk factors for heart disease in healthy subjects." *The Journal of Nutrition* 128 (3): 593-597.

Cooper-Driver, G. A. 2001. "Contributions of Jeffrey Harborne and co-workers to the study of anthocyanins." *Phytochemistry* 56 (3): 229-236.

Costa, M. F., Pimentel, T. C., Guimaraes, J. T., Balthazar, C. F., Rocha, R. S., Cavalcanti, R. N., Esmerino, E. A., Freitas, M. Q., Raices, R. S., and Silva, M. C. 2019. "Impact of prebiotics on the rheological characteristics and volatile compounds of Greek yogurt." *LWT* 105: 371-376.

Creasy, L., and Coffee, M. 1988. "Phytoalexin production potential of grape berries." *Journal of the American Society for Horticultural Science* 113 (2): 230-234.

Crovesy, L., El-Bacha, T., and Rosado, E. L. 2021. "Modulation of the gut microbiota by probiotics and symbiotics is associated with changes in serum metabolite profile related to a decrease in inflammation and overall benefits to metabolic health: a double-blind randomized controlled clinical trial in women with obesity." *Food & Function*.

Cummings, J. H., Antoine, J.-M., Azpiroz, F., Bourdet-Sicard, R., Brandtzaeg, P., Calder, P. C., Gibson, G. R., Guarner, F., Isolauri, E., and Pannemans, D. 2004. "Gut health and immunity." *European Journal of Nutrition* 43:118-173.

Daglia, M. 2012. "Polyphenols as antimicrobial agents." *Current Opinion in Biotechnology* 23 (2): 174-181.

Daliri, E. B.-M., Ofosu, F. K., Chelliah, R., Park, M. H., Kim, J.-H., and Oh, D.-H. 2019. "Development of a soy protein hydrolysate with an antihypertensive effect." *International Journal of Molecular Sciences* 20 (6): 1496.

de Barcelos, S. C., do Egito, A. S., dos Santos, K. M. O., de Moraes, G. M. D., and Teixeira Sá, D. M. A. 2020. "Effect of acerola (Malpighia emarginata DC) pulp incorporation on potentially probiotic Petit-Suisse goat cheese." *Journal of Food Processing and Preservation* 44 (7): e14511.

de Campo, C., Assis, R. Q., da Silva, M. M., Costa, T. M. H., Paese, K., Guterres, S. S., de Oliveira Rios, A., and Flôres, S. H. 2019. "Incorporation of zeaxanthin nanoparticles in yogurt: Influence on physicochemical properties, carotenoid stability and sensory analysis." *Food Chemistry* 301: 125230.

de Mattos, L. L., and Martins, I. S. 2000. "Consumo de fibras alimentares em população adulta." ["Consumption of dietary fibres in adult population"]. *Revista de Saúde Pública* 34 (1): 50-55.

de Moura, S. C., Schettini, G. N., Garcia, A. O., Gallina, D. A., Alvim, I. D., and Hubinger, M. D. 2019. "Stability of hibiscus extract encapsulated by ionic gelation incorporated in yogurt." *Food and Bioprocess Technology* 12 (9): 1500-1515.

de Vries, J. H., Jansen, A., Kromhout, D., van de Bovenkamp, P., van Staveren, W. A., Mensink, R. P., and Katan, M. B. 1997. "The fatty acid and sterol content of food composites of middle-aged men in seven countries." *Journal of Food Composition and Analysis* 10 (2): 115-141.

DeFelice, S. 2002. FIM Rationale and Proposed Guidelines for the Nutraceutical Research & Education Act-NREA, November 10, 2002. Foundation for Innovation in Medicine. www.fimdefelice.org/archives/arc.researchact.

Del Rio, D., Calani, L., Cordero, C., Salvatore, S., Pellegrini, N., and Brighenti, F. 2010. "Bioavailability and catabolism of green tea flavan-3-ols in humans." *Nutrition* 26 (11): 1110-1116.

Delzenne, N. M., Neyrinck, A. M., and Cani, P. D. 2013. "Gut microbiota and metabolic disorders: how prebiotic can work?" *British Journal of Nutrition* 109 (S2): S81-S85.

Depeint, F., Gee, J., Williamson, G., and Johnson, I. 2002. "Evidence for consistent patterns between flavonoid structures and cellular activities." *Proceedings of the Nutrition Society* 61 (1): 97-103.

Dhingra, D., Michael, M., Rajput, H., and Patil, R. 2012. "Dietary fibre in foods: a review." *Journal of Food Science and Technology* 49 (3): 255-266.

Di Mascio, P., Kaiser, S., and Sies, H. 1989. "Lycopene as the most efficient biological carotenoid singlet oxygen quencher." *Archives of Biochemistry and Biophysics* 274 (2): 532-538.

Ding, J., Liang, R., Yang, Y., Sun, N., and Lin, S. 2020. "Optimization of pea protein hydrolysate preparation and purification of antioxidant peptides based on an in silico analytical approach." *LWT* 123: 109126.

Diplock, A. T. 1991. "Antioxidant nutrients and disease prevention: an overview." *The American Journal of Clinical Nutrition* 53 (1 Suppl): 189S-193S.

Duseja, A., Acharya, S. K., Mehta, M., Chhabra, S., Rana, S., Das, A., Dattagupta, S., Dhiman, R. K., and Chawla, Y. K. 2019. "High potency multistrain probiotic improves liver histology in non-alcoholic fatty liver disease (NAFLD): a randomised, double-blind, proof of concept study." *BMJ open gastroenterology* 6 (1): e000315.

Eberhart, C. E., Coffey, R. J., Radhika, A., Giardiello, F. M., Ferrenbach, S., and Dubois, R. N. 1994. "Up-regulation of cyclooxygenase 2 gene expression in human colorectal adenomas and adenocarcinomas." *Gastroenterology* 107 (4): 1183-1188.

Edwards-Jones, G., i Canals, L. M., Hounsome, N., Truninger, M., Koerber, G., Hounsome, B., Cross, P., York, E. H., Hospido, A., and Plassmann, K. 2008. "Testing the assertion that 'local food is best': the challenges of an evidence-based approach." *Trends in Food Science & Technology* 19 (5): 265-274.

Eggersdorfer, M., and Wyss, A. 2018. "Carotenoids in human nutrition and health." *Archives of Biochemistry and Biophysics* 652: 18-26.

Ellegård, L. H., Andersson, S. W., Normén, A. L., and Andersson, H. A. 2007. "Dietary plant sterols and cholesterol metabolism." *Nutrition Reviews* 65 (1): 39-45.

Elsebaie, E. M., and Essa, R. Y. 2018. "Microencapsulation of red onion peel polyphenols fractions by freeze drying technicality and its application in cake." *Journal of Food Processing and Preservation* 42 (7): e13654.

Emami, N., Dehghan, P., Mohtarami, F., Ostadrahimi, A., and Azizi, M. 2018. "Physicochemical, Textural, and Sensory Evaluation of Reduced Fat Gluten-Free Biscuit Prepared with Inulin and Resistant Dextrin Prebiotic." *Journal of Agricultural Science and Technology* 20 (4): 719-731.

Emim, J. A. D. S., Oliveira, A. B., and Lapa, A. J. 1994. "Pharmacological Evaluation of the Anti-inflammatory Activity of a Citrus Bioflavonoid, Hesperidin, and the Isoflavonoids, Duartin and Claussequinone, in Rats and Mice." *Journal of Pharmacy and Pharmacology* 46 (2): 118-122.

Epstein, J., Sanderson, I. R., and MacDonald, T. T. 2010. "Curcumin as a therapeutic agent: the evidence from in vitro, animal and human studies." *British Journal of Nutrition* 103 (11): 1545-1557.

Farzaei, H., M., Rahimi, R., and Abdollahi, M. 2015. "The role of dietary polyphenols in the management of inflammatory bowel disease." *Current Pharmaceutical Biotechnology* 16 (3): 196-210.

Feng, R., Lu, Y., Bowman, L. L., Qian, Y., Castranova, V., and Ding, M. 2005. "Inhibition of activator protein-1, NF-κB, and MAPKs and induction of phase 2 detoxifying enzyme activity by chlorogenic acid." *Journal of Biological Chemistry* 280 (30): 27888-27895.

Fitzgerald, R. J., and Murray, B. A. 2006. "Bioactive peptides and lactic fermentations." *International Journal of Dairy Technology* 59 (2): 118-125.

Foltz, M., Ansems, P., Schwarz, J., Tasker, M. C., Lourbakos, A., and Gerhardt, C. C. 2008. "Protein hydrolysates induce CCK release from enteroendocrine cells and act as partial agonists of the CCK1 receptor." *Journal of Agricultural and Food Chemistry* 56 (3): 837-843.

Fontana, L., Bermudez-Brito, M., Plaza-Diaz, J., Munoz-Quezada, S., and Gil, A. 2013. "Sources, isolation, characterisation and evaluation of probiotics." *British Journal of Nutrition* 109 (S2): S35-S50.

Frémont, L. 2000. "Biological effects of resveratrol." *Life Sciences* 66 (8): 663-673.

Fritz, H., Seely, D., Flower, G., Skidmore, B., Fernandes, R., Vadeboncoeur, S., Kennedy, D., Cooley, K., Wong, R., and Sagar, S. 2013. "Soy, red clover, and isoflavones and breast cancer: a systematic review." *PloS one* 8 (11): e81968.

Gaglio, R., Restivo, I., Barbera, M., Barbaccia, P., Ponte, M., Tesoriere, L., Bonanno, A., Attanzio, A., Di Grigoli, A., and Francesca, N. 2021. "Effect on the Antioxidant, Lipoperoxyl Radical Scavenger Capacity, Nutritional, Sensory and Microbiological Traits of an Ovine Stretched Cheese Produced with Grape Pomace Powder Addition." *Antioxidants* 10 (2): 306.

Gao, X., Chen, Y., Chen, Z., Xue, Z., Jia, Y., Ma, Q., Zhang, M., and Chen, H. 2019. "Identification and antimicrobial activity evaluation of three peptides from laba garlic and the related mechanism." *Food & Function* 10 (8): 4486-4496.

Garcia-Llatas, G., Cilla, A., Alegría, A., and Lagarda, M. J. 2015. "Bioavailability of plant sterol-enriched milk-based fruit beverages: in vivo and in vitro studies." *Journal of Functional Foods* 14: 44-50.

Garcia-Santos, M. D. S. L., Conceicao, F. S., Villas Boas, F., Salotti De Souza, B. M., and Barretto, A. C. D. S. 2019. "Effect of the addition of resistant starch in sausage with fat reduction on the physicochemical and sensory properties." *Food Science and Technology* 39: 491-497.

Gebauer, S. K., Psota, T. L., Harris, W. S., and Kris-Etherton, P. M. 2006. "n−3 Fatty acid dietary recommendations and food sources to achieve essentiality and cardiovascular benefits." *The American Journal of Clinical Nutrition* 83 (6): S1526-1535S.

Ghafouri-Oskuei, H., Javadi, A., Asl, M. R. S., Azadmard-Damirchi, S., and Armin, M. 2020. "Quality properties of sausage incorporated with flaxseed and tomato powders." *Meat Science* 161: 107957.

Gibson, G. R., and Roberfroid, M. B. 1995. "Dietary modulation of the human colonic microbiota: introducing the concept of prebiotics." *The Journal of Nutrition* 125 (6): 1401.

Gilani, A. H. 2005. "Trends in ethnopharmacology." *Journal of Ethnopharmaco-logy* 100 (1): 43-49.

Girgih, A. T., He, R., Malomo, S., Offengenden, M., Wu, J., and Aluko, R. E. 2014. "Structural and functional characterization of hemp seed (Cannabis sativa L.) protein-derived antioxidant and antihypertensive peptides." *Journal of Functional Foods* 6: 384-394.

Gobbetti, M., Minervini, F., and Rizzello, C. G. 2007. "Bioactive peptides in dairy products." *Handbook of food products manufacturing*: 489-517.

Gong, K., Deng, L., Shi, A., Liu, H., Liu, L., Hu, H., Adhikari, B., and Wang, Q. 2017. "High-pressure microfluidisation pretreatment disaggregate peanut protein isolates to prepare antihypertensive peptide fractions." *International Journal of Food Science & Technology* 52 (8): 1760-1769.

Graham, H. N. 1992. "Green tea composition, consumption, and polyphenol chemistry." *Preventive Medicine* 21 (3): 334-350.

Grasso, S., Brunton, N. P., Lyng, J. G., Harrison, S. M., and Monahan, F. J. 2016. "Quality of deli-style turkey enriched with plant sterols." *Food Science and Technology International* 22 (8): 743-751.

Gross, J. 1991. "Carotenoids." In *Pigments in Vegetables*, 75-278. Springer.

Grosso, G., Galvano, F., Mistretta, A., Marventano, S., Nolfo, F., Calabrese, G., Buscemi, S., Drago, F., Veronesi, U., and Scuderi, A. 2013. "Red orange: experimental models and epidemiological evidence of its benefits on human health." *Oxidative Medicine and Cellular Longevity* 2013.

Grosso, G., Stepaniak, U., Micek, A., Topor-Mądry, R., Pikhart, H., Szafraniec, K., and Pająk, A. 2015. "Association of daily coffee and tea consumption and metabolic syndrome: results from the Polish arm of the HAPIEE study." *European Journal of Nutrition* 54 (7): 1129-1137.

Guimarães, J. T., Silva, E. K., Costa, A. L. R., Cunha, R. L., Freitas, M. Q., Meireles, M. A. A., and Cruz, A. G. 2018. "Manufacturing a prebiotic whey beverage exploring the influence of degree of inulin polymerization." *Food Hydrocolloids* 77: 787-795.

Guo, H., Hao, Y., Richel, A., Everaert, N., Chen, Y., Liu, M., Yang, X., and Ren, G. 2020. "Antihypertensive effect of quinoa protein under simulated gastrointestinal digestion and peptide characterization." *Journal of the Science of Food and Agriculture* 100 (15): 5569-5576.

Halliwell, B., Rafter, J., and Jenner, A. 2005. "Health promotion by flavonoids, tocopherols, tocotrienols, and other phenols: direct or indirect effects? Antioxidant or not?" *The American Journal of Clinical Nutrition* 81 (1): 268S-276S.

Han, L., Zhang, J., and Cao, X. 2021. "Effects of orange peel powder on rheological properties of wheat dough and bread aging." *Food Science & Nutrition* 9 (2): 1061-1069.

Han, R.-M., Tian, Y.-X., Liu, Y., Chen, C.-H., Ai, X.-C., Zhang, J.-P., and Skibsted, L. H. 2009. "Comparison of flavonoids and isoflavonoids as antioxidants." *Journal of Agricultural and Food Chemistry* 57 (9): 3780-3785.

Harborne, J. B., and Williams, C. A. 2000. "Advances in flavonoid research since 1992." *Phytochemistry* 55 (6): 481-504.

Harnedy, P. A., Parthsarathy, V., McLaughlin, C. M., O'Keeffe, M. B., Allsopp, P. J., McSorley, E. M., O'Harte, F. P., and FitzGerald, R. J. 2018. "Atlantic salmon (Salmo salar) co-product-derived protein hydrolysates: A source of antidiabetic peptides." *Food Research International* 106: 598-606.

Hartmann, R., and Meisel, H. 2007. "Food-derived peptides with biological activity: from research to food applications." *Current Opinion in Biotechnology* 18 (2): 163-169.

Hassellund, S., Flaa, A., Kjeldsen, S., Seljeflot, I., Karlsen, A., Erlund, I., and Rostrup, M. 2013. "Effects of anthocyanins on cardiovascular risk factors and inflammation in pre-hypertensive men: a double-blind randomized placebo-controlled crossover study." *Journal of Human Hypertension* 27 (2): 100-106.

Havsteen, B. H. 2002. "The biochemistry and medical significance of the flavonoids." *Pharmacology & Therapeutics* 96 (2): 67-202.

Hayes, M., Carney, B., Slater, J., and Brück, W. 2008a. "Mining marine shellfish wastes for bioactive molecules: Chitin and chitosan–Part B: Applications." *Biotechnology Journal* 3 (7): 878-889.

Hayes, M. V., J. 2013. "Nutraceuticals derived from marine sources: Extraction and application of bioactive carbohydrates, proteins and peptides as functional ingredients." In *Nutraceutical and Functional Foods: Natural Remedy*, edited by S. K. Brar, Kaur, S., Dhillon, G. S., 39-49. NRS: Québec, QC, Canada.

Hempel, S., Newberry, S. J., Maher, A. R., Wang, Z., Miles, J. N., Shanman, R., Johnsen, B., and Shekelle, P. G. 2012. "Probiotics for the prevention and treatment of antibiotic-associated diarrhea: a systematic review and meta-analysis." *Jama* 307 (18): 1959-1969.

Heydari, S., Amiri-Rigi, A., Ehsani, M. R., Mohammadifar, M. A., Khorshidian, N., Koushki, M. R., and Mortazavian, A. M. 2018. "Rheological behaviour, sensory properties and syneresis of probiotic yoghurt supplemented with various prebiotics." *International Journal of Dairy Technology* 71: 175-184.

Hicks, K. B., and Moreau, R. A. 2001. "Phytosterols and phytostanols: functional food cholesterol busters." *Food Technology* 55 (1): 63-67.

Hill, C., Guarner, F., Reid, G., Gibson, G. R., Merenstein, D. J., Pot, B., Morelli, L., Canani, R. B., Flint, H. J., and Salminen, S. 2014. "Expert consensus document: The International Scientific Association for Probiotics and Prebiotics consensus statement on the scope and appropriate use of the term probiotic." *Nature Reviews Gastroenterology & Hepatology* 11 (8): 506-514.

Hilz, H., Bakx, E. J., Schols, H. A., and Voragen, A. G. 2005. "Cell wall polysaccharides in black currants and bilberries—characterisation in berries, juice, and press cake." *Carbohydrate Polymers* 59 (4): 477-488.

Hirata, A., Murakami, Y., Shoji, M., Kadoma, Y., And Fujisawa, S. 2005. "Kinetics of radical-scavenging activity of hesperetin and hesperidin and their inhibitory activity on COX-2 expression." *Anticancer Research* 25 (5): 3367-3374.

Hosseinvand, A., Sohrabvandi, S., Yousefi, M., Khorshidian, N., and Khoshtinat, K. 2019. "Effect of corn bran as a fat replacer on quality characteristics of low-fat mayonnaise sauce." *Biointerface Research in Applied Chemistry* 9 (5): 4248-4254.

Hotel, A. C. P., and Cordoba, A. 2001. "Health and nutritional properties of probiotics in food including powder milk with live lactic acid bacteria." *Prevention* 5 (1).

Hu, F., Ci, A.-T., Wang, H., Zhang, Y.-Y., Zhang, J.-G., Thakur, K., and Wei, Z.-J. 2018. "Identification and hydrolysis kinetic of a novel antioxidant peptide from pecan meal using Alcalase." *Food Chemistry* 261: 301-310.

Hu, S., Fan, X., Qi, P., and Zhang, X. 2019. "Identification of anti-diabetes peptides from Spirulina platensis." *Journal of Functional Foods* 56: 333-341.

Hügel, H. M., Jackson, N., May, B., Zhang, A. L., and Xue, C. C. 2016. "Polyphenol protection and treatment of hypertension." *Phytomedicine* 23 (2): 220-231.

Isolauri, E., Rautava, S., and Salminen, S. 2012. "Probiotics in the development and treatment of allergic disease." *Gastroenterology clinics of North America* 41 (4): 747-762.

Izadi, Z., Nasirpour, A., Garoosi, G. A., and Tamjidi, F. 2015. "Rheological and physical properties of yogurt enriched with phytosterol during storage." *Journal of Food Science and Technology* 52 (8): 5341-5346.

Jackson, C.-J., Dini, J., Lavandier, C., Rupasinghe, H., Faulkner, H., Poysa, V., Buzzell, D., and DeGrandis, S. 2002. "Effects of processing on the content and composition of isoflavones during manufacturing of soy beverage and tofu." *Process Biochemistry* 37 (10): 1117-1123.

Jagetia, G. C., and Aggarwal, B. B. 2007. ""Spicing up" of the immune system by curcumin." *Journal of Clinical Immunology* 27 (1): 19-35.

Jayawardana, B. C., Warnasooriya, V. B., Thotawattage, G., Dharmasena, V., and Liyanage, R. 2019. "Black and green tea (Camellia sinensis L.) extracts as natural antioxidants in uncured pork sausages." *Journal of Food Processing and Preservation* 43 (2): e13870.

Jeandet, P., Bessis, R., and Gautheron, B. 1991. "The production of resveratrol (3, 5, 4'-trihydroxystilbene) by grape berries in different developmental stages." *American Journal of Enology and Viticulture* 42 (1): 41-46.

Jiang, X., Cui, Z., Wang, L., Xu, H., and Zhang, Y. 2020. "Production of bioactive peptides from corn gluten meal by solid-state fermentation with Bacillus subtilis MTCC5480 and evaluation of its antioxidant capacity in vivo." *LWT* 131: 109767.

Johnson, E. J. 2002. "The role of carotenoids in human health." *Nutrition in Clinical Care* 5 (2): 56-65.

Jokura, H., Watanabe, I., Umeda, M., Hase, T., and Shimotoyodome, A. 2015. "Coffee polyphenol consumption improves postprandial hyperglycemia associated with impaired vascular endothelial function in healthy male adults." *Nutrition Research* 35 (10): 873-881.

Jovanovic, S., Steenken, S., Simic, M., and Hara, Y. 1998. "Antioxidant properties of flavonoids: reduction potentials and electron transfer reactions of flavonoid radicals." *Marcel Dekker*: 137-161.

Jovanovic, S. V., and Simic, M. G. 2000. "Antioxidants in nutrition." *Annals of the New York Academy of Sciences* 899 (1): 326-334.

Jurenka, J. S. 2009. "Anti-inflammatory properties of curcumin, a major constituent of Curcuma longa: a review of preclinical and clinical research." *Alternative Medicine Review* 14 (2).

Kabara, J. J. 2000. "Health oils from the tree of life." *Nutritional and Health Aspects of Coconut Oil). Indian Coconut Journal* 31 (8): 2-8.

Karimi, A., Azizi, M. H., and Ahmadi Gavlighi, H. 2020. "Fractionation of hydrolysate from corn germ protein by ultrafiltration: In vitro antidiabetic and antioxidant activity." *Food Science & Nutrition* 8 (5): 2395-2405.

Kaulmann, A., and Bohn, T. 2014. "Carotenoids, inflammation, and oxidative stress—implications of cellular signaling pathways and relation to chronic disease prevention." *Nutrition Research* 34 (11): 907-929.

Kaur, R., and Riar, C. S. 2020. "Sensory, rheological and chemical characteristics during storage of set type full fat yoghurt fortified with barley β-glucan." *Journal of Food Science and Technology* 57 (1): 41-51.

Kerry, R. G., Patra, J. K., Gouda, S., Park, Y., Shin, H.-S., and Das, G. 2018. "Benefaction of probiotics for human health: A review." *Journal of Food and Drug Analysis* 26 (3): 927-939.

Khachik, F., Beecher, G. R., and Smith, J. C. 1995. "Lutein, lycopene, and their oxidative metabolites in chemoprevention of cancer." *Journal of Cellular Biochemistry* 59 (S22): 236-246.

Khachik, F., Carvalho, L., Bernstein, P. S., Muir, G. J., Zhao, D.-Y., and Katz, N. B. 2002. "Chemistry, distribution, and metabolism of tomato carotenoids and their impact on human health." *Experimental Biology and Medicine* 227 (10): 845-851.

Khan, B. M., Qiu, H.-M., Xu, S.-Y., Liu, Y., and Cheong, K.-L. 2020. "Physicochemical characterization and antioxidant activity of sulphated polysaccharides derived from Porphyra haitanensis." *International Journal of Biological Macromolecules* 145: 1155-1161.

Khorshidian, N., Yousefi, M., Shadnoush, M., and Mortazavian, A. M. 2018. "An overview of β-Glucan functionality in dairy products." *Current Nutrition & Food Science* 14 (4): 280-292.

Khorshidian, N., Yousefi, M., Shadnoush, M., Siadat, S. D., Mohammadi, M., and Mortazavian, A. M. 2020. "Using probiotics for mitigation of acrylamide in food products: a mini review." *Current Opinion in Food Science* 32: 67-75.

Kim, S. J., and Kim, H. M. 2017. "Dynamic lipopolysaccharide transfer cascade to TLR4/MD2 complex via LBP and CD14." *BMB Reports* 50 (2): 55.

Kitts, D. D., and Weiler, K. 2003. "Bioactive proteins and peptides from food sources. Applications of bioprocesses used in isolation and recovery." *Current Pharmaceutical Design* 9 (16): 1309-1323.

Knekt, P., Kumpulainen, J., Järvinen, R., Rissanen, H., Heliövaara, M., Reunanen, A., Hakulinen, T., and Aromaa, A. 2002. "Flavonoid intake and risk of chronic diseases." *The American Journal of Clinical Nutrition* 76 (3): 560-568.

Krinsky, N. I., and Johnson, E. J. 2005. "Carotenoid actions and their relation to health and disease." *Molecular Aspects of Medicine* 26 (6): 459-516.

Kris-Etherton, P., Taylor, D. S., Yu-Poth, S., Huth, P., Moriarty, K., Fishell, V., Hargrove, R. L., Zhao, G., and Etherton, T. D. 2000. "Polyunsaturated fatty acids in the food chain in the United States." *The American Journal of Clinical Nutrition* 71 (1): 179S-188S.

Kris-Etherton, P. M., Innis, S., and Ammerican, D. A. 2007. "Position of the American Dietetic Association and Dietitians of Canada: dietary fatty acids." *Journal of the American Dietetic Association* 107 (9): 1599-1611.

Krumbeck, J. A., Walter, J., and Hutkins, R. W. 2018. "Synbiotics for improved human health: recent developments, challenges, and opportunities." *Annual Review of Food Science and Technology* 9: 451-479.

Kuhnle, G. G., Dell'Aquila, C., Aspinall, S. M., Runswick, S. A., Mulligan, A. A., and Bingham, S. A. 2008. "Phytoestrogen content of foods of animal origin: dairy products, eggs, meat, fish, and seafood." *Journal of Agricultural and Food Chemistry* 56 (21): 10099-10104.

Kulthe, A., Thorat, S., and Lande, S. 2017. "Preparation of β-Carotene Enriched Pearl Millet Based Cookies." *International Journal of Current Microbiology and Applied Sciences* 6 (2): 1197-1203.

Kumar, A., Krishnamoorthy, E., Devi, H. M., Uchoi, D., Tejpal, C., Ninan, G., and Zynudheen, A. 2018. "Influence of sea grapes (Caulerpa racemosa) supplementation on physical, functional, and anti-oxidant properties of semi-sweet biscuits." *Journal of Applied Phycology* 30 (2): 1393-1403.

La Casa, C., Villegas, I., De La Lastra, C. A., Motilva, V., and Calero, M. M. 2000. "Evidence for protective and antioxidant properties of rutin, a natural flavone, against ethanol induced gastric lesions." *Journal of Ethnopharmacology* 71 (1): 45-53.

Lairon, D., Arnault, N., Bertrais, S., Planells, R., Clero, E., Hercberg, S., and Boutron-Ruault, M.-C. 2005. "Dietary fiber intake and risk factors for cardiovascular disease in French adults." *The American Journal of Clinical Nutrition* 82 (6): 1185-1194.

Lamothe, S., Langlois, A., Bazinet, L., Couillard, C., and Britten, M. 2016. "Antioxidant activity and nutrient release from polyphenol-enriched cheese in a simulated gastrointestinal environment." *Food & Function* 7 (3): 1634-1644.

Laparra, J. M., and Sanz, Y. 2010. "Interactions of gut microbiota with functional food components and nutraceuticals." *Pharmacological Research* 61 (3): 219-225.

Laranjinha, J. A., Almeida, L. M., and Madeira, V. M. 1994. "Reactivity of dietary phenolic acids with peroxyl radicals: antioxidant activity upon low density lipoprotein peroxidation." *Biochemical Pharmacology* 48 (3): 487-494.

Lattimer, J. M., and Haub, M. D. 2010. "Effects of dietary fiber and its components on metabolic health." *Nutrients* 2 (12): 1266-1289.

Lee, J., Koo, N., and Min, D. B. 2004. "Reactive oxygen species, aging, and antioxidative nutraceuticals." *Comprehensive Reviews in Food Science and Food Safety* 3 (1): 21-33.

Lee, T.-C., and Ho, C.-T. 2002. *Bioactive compounds in foods: effects of processing and storage.* ACS Publications.

Li, D., Zhang, Y., Liu, Y., Sun, R., and Xia, M. 2015. "Purified anthocyanin supplementation reduces dyslipidemia, enhances antioxidant capacity, and prevents insulin resistance in diabetic patients." *The Journal of Nutrition* 145 (4): 742-748.

Liao, W., Fan, H., and Wu, J. 2018. "Egg white-derived antihypertensive peptide IRW (Ile-Arg-Trp) inhibits angiotensin II-stimulated migration of vascular smooth muscle cells via angiotensin type I receptor." *Journal of Agricultural and Food Chemistry* 66 (20): 5133-5138.

Lichtenstein, A. H., Appel, L. J., Brands, M., Carnethon, M., Daniels, S., Franch, H. A., Franklin, B., Kris-Etherton, P., Harris, W. S., and Howard, B. 2006. "Diet and lifestyle recommendations revision 2006." *Circulation* 114 (1): 82-96.

Lima, J. T., Almeida, J. R., Barbosa-Filho, J. M., Assis, T. S., Silva, M. S., da-Cunha, E. V., Braz-Filhod, R., and Silva, B. A. 2005. "Spasmolytic action of diplotropin, a furanoflavan from Diplotropis ferruginea Benth., involves calcium blockade in guinea-pig ileum." *Zeitschrift für Naturforschung B* 60 (10): 1093-1100.

Ling, W., and Jones, P. 1995. "Enhanced efficacy of sitostanol-containing versus sitostanol-free phytosterol mixtures in altering lipoprotein cholesterol levels and synthesis in rats." *Atherosclerosis* 118 (2): 319-331.

Liu, J., Willför, S., and Xu, C. 2015. "A review of bioactive plant polysaccharides: Biological activities, functionalization, and biomedical applications." *Bioactive Carbohydrates and Dietary Fibre* 5 (1): 31-61.

Liu, K. 1997. "Chemistry and nutritional value of soybean components." In *Soybeans*, 25-113. Springer.

Liu, K., Du, R., and Chen, F. 2020. "Stability of the antioxidant peptide SeMet-Pro-Ser identified from selenized brown rice protein hydrolysates." *Food Chemistry* 319: 126540.

Liu, S., Stampfer, M. J., Hu, F. B., Giovannucci, E., Rimm, E., Manson, J. E., Hennekens, C. H., and Willett, W. C. 1999. "Whole-grain consumption and risk of coronary heart disease: results from the Nurses' Health Study." *The American Journal of Clinical Nutrition* 70 (3): 412-419.

Lordan, S., Ross, R. P., and Stanton, C. 2011. "Marine bioactives as functional food ingredients: potential to reduce the incidence of chronic diseases." *Marine drugs* 9 (6): 1056-1100.

Lu, X., Zhang, L., Sun, Q., Song, G., and Huang, J. 2019. "Extraction, identification and structure-activity relationship of antioxidant peptides from sesame (Sesamum indicum L.) protein hydrolysate." *Food Research International* 116: 707-716.

Ma, Z., Zhang, C., Gao, X., Cui, F., Zhang, J., Jia, M., Jia, S., and Jia, L. 2015. "Enzymatic and acidic degradation effect on intracellular polysaccharide of Flammulina velutipes SF-08." *International Journal of Biological Macromolecules* 73: 236-244.

Magrone, T., Candore, G., Caruso, C., Jirillo, E., and Covelli, V. 2008. "Polyphenols from red wine modulate immune responsiveness: biological and clinical significance." *Current Pharmaceutical Design* 14 (26): 2733-2748.

Maiani, G., Periago Castón, M. J., Catasta, G., Toti, E., Cambrodón, I. G., Bysted, A., Granado-Lorencio, F., Olmedilla-Alonso, B., Knuthsen, P., and Valoti, M. 2009. "Carotenoids: actual knowledge on food sources, intakes, stability and bioavailability and their protective role in humans." *Molecular Nutrition & Food Research* 53 (S2).

Malaguarnera, G., Giordano, M., Nunnari, G., Bertino, G., and Malaguarnera, M. 2014. "Gut microbiota in alcoholic liver disease: pathogenetic role and therapeutic perspectives." *World Journal of Gastroenterology: WJG* 20 (44): 16639.

Manach, C., Morand, C., Gil-Izquierdo, A., Bouteloup-Demange, C., and Remesy, C. 2003. "Bioavailability in humans of the flavanones hesperidin and narirutin after the ingestion of two doses of orange juice." *European Journal of Clinical Nutrition* 57 (2): 235-242.

Manach, C., Scalbert, A., Morand, C., Rémésy, C., and Jiménez, L. 2004. "Polyphenols: food sources and bioavailability." *The American Journal of Clinical Nutrition* 79 (5): 727-747.

Manach, C., Williamson, G., Morand, C., Scalbert, A., and Rémésy, C. 2005. "Bioavailability and bioefficacy of polyphenols in humans. I. Review of 97 bioavailability studies." *The American Journal of Clinical Nutrition* 81 (1): 230S-242S.

Mangels, A. R., Holden, J. M., Beecher, G. R., Forman, M. R., and Lanza, E. 1993. "Carotenoid content of fruits and vegetables: an evaluation of analytic data." *Journal of the American Dietetic Association* 93 (3): 284-296.

Manning, T. S., and Gibson, G. R. 2004. "Prebiotics." *Best Practice & Research Clinical Gastroenterology* 18 (2): 287-298.

Marinelli, V., Padalino, L., Nardiello, D., Del Nobile, M. A., and Conte, A. 2015. "New approach to enrich pasta with polyphenols from grape marc." *Journal of Chemistry* 2015.

Marteau, P., and Shanahan, F. 2003. "Basic aspects and pharmacology of probiotics: an overview of pharmacokinetics, mechanisms of action and side-effects." *Best Practice & Research Clinical Gastroenterology* 17 (5): 725-740.

Marthandam Asokan, S., Wang, T., Su, W.-T., and Lin, W.-T. 2019. "Antidiabetic effects of a short peptide of potato protein hydrolysate in STZ-induced diabetic mice." *Nutrients* 11 (4): 779.

Martin, D. A., and Bolling, B. W. 2015. "A review of the efficacy of dietary polyphenols in experimental models of inflammatory bowel diseases." *Food & Function* 6 (6): 1773-1786.

Martins, C., Da Silva, D., Neres, A., Magalhaes, T., Watanabe, G., Modolo, L., Sabino, A., De Fátima, A., and De Resende, M. 2008. "Curcumin as a promising antifungal of clinical interest." *Journal of Antimicrobial Chemotherapy* 63 (2): 337-339.

Masmoudi, M., Ammar, I., Ghribi, H., and Attia, H. 2020. "Physicochemical, radical scavenging activity and sensory properties of a soft cheese fortified with Arbutus unedo L. extract." *Food Bioscience* 35: 100579.

Matsuzaki, T., and Hara, Y. 1985. "Antioxidative activity of tea leaf catechins." *Nippon Nōgeikagaku Kaishi* 59 (2): 129-134.

Mazorra-Manzano, M. A., Robles-Porchas, G. R., Gonzalez-Velazquez, D. A., Torres-Llanez, M. J., Martínez-Porchas, M., García-Sifuentes, C. O., González-Córdova, A. F., and Vallejo-Cordoba, B. 2020. "Cheese whey fermentation by its native microbiota: Proteolysis and bioactive peptides release with ACE-inhibitory activity." *Fermentation* 6: 19.

Mazza, G. 1998. *Functional Foods: Biochemical and Processing Aspects*. Vol. 1. CRC Press.

McFarland, L. V. 2007. "Meta-analysis of probiotics for the prevention of traveler's diarrhea." *Travel Medicine and Infectious Disease* 5 (2): 97-105.

McGuire, S. 2011. "US Department of Agriculture and US Department of Health and Human Services, Dietary Guidelines for Americans, 2010. Washington, DC: US Government Printing Office, January 2011." *Advances in Nutrition: An International Review Journal* 2 (3): 293-294.

Megrous, S., Al-Dalali, S., Zhao, X., Chen, C., Cao, Y., Bourouis, I., Mekkaoui, A., Yang, Z., and Yang, Z. 2020. "Evaluation of Antidiabetic Activities of Casein Hydrolysates by a Bacillus Metalloendopeptidase." *International Journal of Peptide Research and Therapeutics*: 1-9.

Mellor, D., Madden, L., Smith, K., Kilpatrick, E., and Atkin, S. 2013. "High-polyphenol chocolate reduces endothelial dysfunction and oxidative stress during acute transient hyperglycaemia in Type 2 diabetes: a pilot randomized controlled trial." *Diabetic Medicine* 30 (4): 478-483.

Messina, M., Barnes, S., and Setchell, K. D. 1997. "Phyto-oestrogens and breast cancer." *The Lancet* 350 (9083): 971-972.

Messina, M. J. 1997. "Soyfoods: Their role in disease prevention and treatment." In *Soybeans*, 442-477. Springer.

Meyer, D. 2015. "Health benefits of prebiotic fibers." *Advances in Food and Nutrition Research* 74: 47-91.

Michael, D., Jack, A., Masetti, G., Davies, T., Loxley, K., Kerry-Smith, J., Plummer, J., Marchesi, J., Mullish, B., and McDonald, J. 2020. "A randomised controlled study shows supplementation of overweight and obese adults with lactobacilli and bifidobacteria reduces bodyweight and improves well-being." *Scientific Reports* 10 (1): 1-12.

Miettinen, T., and Gylling, H. 2004. "Plant stanol and sterol esters in prevention of cardiovascular diseases." *Annals of Medicine* 36 (2): 126-134.

Milenkovic, D., Deval, C., Dubray, C., Mazur, A., and Morand, C. 2011. "Hesperidin displays relevant role in the nutrigenomic effect of orange juice on blood leukocytes in human volunteers: a randomized controlled cross-over study." *PLoS One* 6 (11): e26669.

Milner, J. A., Romagnolo, D. F., Connor, J., and Lee, S. 2010. *Bioactive Compounds and Cancer*. Springer.

Mine, Y., Li-Chan, E., and Jiang, B. 2011. *Bioactive Proteins and Peptides as Functional Foods and Nutraceuticals*. Vol. 29. John Wiley & Sons.

Miwa, Y., Mitsuzumi, H., Sunayama, T., Yamada, M., Okada, K., Kubota, M., Chaen, H., Mishima, Y., and Kibata, M. 2005. "Glucosyl hesperidin lowers serum triglyceride level in hypertriglyceridemic subjects through the improvement of very low-density lipoprotein metabolic abnormality." *Journal of Nutritional Science and Vitaminology* 51 (6): 460-470.

Miwa, Y., Yamada, M., Sunayama, T., Mitsuzumi, H., Tsuzaki, Y., Chaen, H., Mishima, Y., and Kibata, M. 2004. "Effects of glucosyl hesperidin on serum lipids in hyperlipidemic subjects: preferential reduction in elevated serum triglyceride level." *Journal of Nutritional Science and Vitaminology* 50 (3): 211-218.

Mohammadi, R., Sohrabvandi, S., and Mohammad Mortazavian, A. 2012. "The starter culture characteristics of probiotic microorganisms in fermented milks." *Engineering in Life Sciences* 12 (4): 399-409.

Mohebbi, Z., Homayouni, A., Azizi, M. H., and Hosseini, S. J. 2018. "Effects of beta-glucan and resistant starch on wheat dough and prebiotic bread properties." *Journal of Food Science and Technology* 55 (1): 101-110.

Mojica, L., and De Mejía, E. G. 2016. "Optimization of enzymatic production of anti-diabetic peptides from black bean (Phaseolus vulgaris L.) proteins, their characterization and biological potential." *Food & Function* 7 (2): 713-727.

Molina, C. V., Lima, J. G., Moraes, I. C., and Pinho, S. C. 2019. "Physicochemical characterization and sensory evaluation of yogurts incorporated with beta-carotene-loaded solid lipid microparticles stabilized with hydrolyzed soy protein isolate." *Food Science and Biotechnology* 28 (1): 59-66.

Monforte, M., Trovato, A., Kirjavainen, S., Forestieri, A., Galati, E., and Lo, R. C. 1995. "Biological effects of hesperidin, a Citrus flavonoid. (note II): hypolipidemic activity on experimental hypercholesterolemia in rat." *Farmaco (SocietÃ chimica italiana: 1989)* 50 (9): 595-599.

Montonen, J., Knekt, P., Järvinen, R., Aromaa, A., and Reunanen, A. 2003. "Whole-grain and fiber intake and the incidence of type 2 diabetes." *The American Journal of Clinical Nutrition* 77 (3): 622-629.

Moreau, R. A. 2003. "Plant Sterols in Functional Foods." *Phytosterols as Functional Food Components and Nutraceuticals*: 317.

Moreau, R. A., Whitaker, B. D., and Hicks, K. B. 2002. "Phytosterols, phytostanols, and their conjugates in foods: structural diversity, quantitative analysis, and health-promoting uses." *Progress in Lipid Research* 41 (6): 457-500.

Morton, G. M., Lee, S. M., Buss, D. H., and Lawrance, P. 1995. "Intakes and major dietary sources of cholesterol and phytosterols in the British diet." *Journal of Human Nutrition and Dietetics* 8 (6): 429-440.

Mudgil, P., Kamal, H., Yuen, G. C., and Maqsood, S. 2018. "Characterization and identification of novel antidiabetic and anti-obesity peptides from camel milk protein hydrolysates." *Food Chemistry* 259: 46-54.

Nagao, T., Ochiai, R., Watanabe, T., Kataoka, K., Komikado, M., Tokimitsu, I., and Tsuchida, T. 2009. "Visceral fat-reducing effect of continuous coffee beverage consumption in obese subjects." *Japanese Pharmacology and Therapeutics* 37 (4): 333-344.

Neves, A. C., Harnedy, P. A., O'Keeffe, M. B., Alashi, M. A., Aluko, R. E., and FitzGerald, R. J. 2017. "Peptide identification in a salmon gelatin hydrolysate with antihypertensive, dipeptidyl peptidase IV inhibitory and antioxidant activities." *Food Research International* 100: 112-120.

Nollet, L. M., and Toldrá, F. 2012. *Handbook of Analysis of Active Compounds in Functional Foods*. CRC Press.

Normén, A. L., Brants, H. A., Voorrips, L. E., Andersson, H. A., van den Brandt, P. A., and Goldbohm, R. A. 2001. "Plant sterol intakes and colorectal cancer risk in the Netherlands Cohort Study on Diet and Cancer." *The American Journal of Clinical Nutrition* 74 (1): 141-148.

Nurdiani, R., Vasiljevic, T., Yeager, T., Singh, T. K., and Donkor, O. N. 2017. "Bioactive peptides with radical scavenging and cancer cell cytotoxic activities derived from Flathead (Platycephalus fuscus) by-products." *European Food Research and Technology* 243 (4): 627-637.

O'Sullivan, L., Murphy, B., McLoughlin, P., Duggan, P., Lawlor, P. G., Hughes, H., and Gardiner, G. E. 2010. "Prebiotics from marine macroalgae for human and animal health applications." *Marine Drugs* 8 (7): 2038-2064.

Oh, J. H., Jang, Y. S., Kang, D., Chang, D. K., and Min, Y. W. 2019. "Efficacy and safety of new Lactobacilli probiotics for unconstipated irritable bowel syndrome: a randomized, double-blind, placebo-controlled trial." *Nutrients* 11 (12): 2887.

Orem, A., Alasalvar, C., Kural, B. V., Yaman, S., Orem, C., Karadag, A., Pelvan, E., and Zawistowski, J. 2017. "Cardio-protective effects of phytosterol-enriched functional black tea in mild hypercholesterolemia subjects." *Journal of Functional Foods* 31: 311-319.

Ota, N., Soga, S., Murase, T., Shimotoyodome, A., and Hase, T. 2010. "Consumption of coffee polyphenols increases fat utilization in humans." *Journal of Health Science* 56 (6): 745-751.

Pal, S., and Bhattacharjee, P. 2019. "Lutein-fortified potato soup and freeze-dried lutein powder designed with supercritical carbon dioxide extract of yellow corn kernels are promising nutraceutical foods." *Journal of Food Processing and Preservation* 43 (8): e14005.

Paliyath, G., Bakovic, M., and Shetty, K. 2011. *Functional Foods, Nutraceuticals and Degenerative Disease Prevention*. John Wiley & Sons.

Paliyath, G., and Shetty, K. 2011. "Functional Foods, Nutraceuticals, and Disease Prevention: A Window to the Future of Health Promotion." *Functional Foods, Nutraceuticals, and Degenerative Disease Prevention*: 3-9.

Palozza, P., Simone, R., Catalano, A., Parrone, N., Monego, G., and Ranelletti, F. O. 2011. "Lycopene regulation of cholesterol synthesis and efflux in human macrophages." *The Journal of Nutritional Biochemistry* 22 (10): 971-978.

Pan, M.-H., and Ho, C.-T. 2008. "Chemopreventive effects of natural dietary compounds on cancer development." *Chemical Society Reviews* 37 (11): 2558-2574.

Pandya, U., Dhuldhaj, U., and Sahay, N. S. 2019. "Bioactive mushroom polysaccharides as antitumor: an overview." *Natural Product Research* 33 (18): 2668-2680.

Panghal, A., Kaur, R., Janghu, S., Sharma, P., Sharma, P., and Chhikara, N. 2019. "Nutritional, phytochemical, functional and sensorial attributes of Syzygium cumini L. pulp incorporated pasta." *Food Chemistry* 289: 723-728.

Panpipat, W., Chaijan, M., and Guo, Z. 2018. "Oxidative stability of margarine enriched with different structures of β-sitosteryl esters during storage." *Food Bioscience* 22: 78-84.

Parnell, J. A., Raman, M., Rioux, K. P., and Reimer, R. A. 2012. "The potential role of prebiotic fibre for treatment and management of non-alcoholic fatty liver disease and associated obesity and insulin resistance." *Liver International* 32 (5): 701-11.

Peterson, J. J., Dwyer, J. T., Beecher, G. R., Bhagwat, S. A., Gebhardt, S. E., Haytowitz, D. B., and Holden, J. M. 2006. "Flavanones in oranges, tangerines (mandarins), tangors, and tangelos: a compilation and review of the data from the analytical literature." *Journal of Food Composition and Analysis* 19: S66-S73.

Petruzziello, L., Iacopini, F., Bulajic, M., Shah, S., and Costamagna, G. 2006. "Uncomplicated diverticular disease of the colon." *Alimentary Pharmacology & Therapeutics* 23 (10): 1379-1391.

Pfeuffer, M., and Schrezenmeir, J. 2002. "Milk lipids in diet and health: medium chain fatty acids (MCFA)." *Bulletin-International Dairy Federation* (377): 32-42.

Piao, M. J., Kim, K. C., Chae, S., Keum, Y. S., Kim, H. S., and Hyun, J. W. 2013. "Protective effect of fisetin (3, 7, 3', 4'-tetrahydroxyflavone) against γ-irradiation-induced oxidative stress and cell damage." *Biomolecules & Therapeutics* 21 (3): 210.

Pietta, P.-G. 2000. "Flavonoids as antioxidants." *Journal of Natural Products* 63 (7): 1035-1042.

Piironen, V., Lindsay, D. G., Miettinen, T. A., Toivo, J., and Lampi, A. M. 2000. "Plant sterols: biosynthesis, biological function and their importance to human nutrition." *Journal of the Science of Food and Agriculture* 80 (7): 939-966.

Pineiro, M., Asp, N.-G., Reid, G., Macfarlane, S., Morelli, L., Brunser, O., and Tuohy, K. 2008. "FAO Technical meeting on prebiotics." *Journal of Clinical Gastroenterology* 42: S156-S159.

Plaza-Diaz, J., Gomez-Llorente, C., Abadia-Molina, F., Saez-Lara, M. J., Campaña-Martin, L., Muñoz-Quezada, S., Romero, F., Gil, A., and Fontana, L. 2014. "Effects of

Lactobacillus paracasei CNCM I-4034, Bifidobacterium breve CNCM I-4035 and Lactobacillus rhamnosus CNCM I-4036 on hepatic steatosis in Zucker rats." *PloS one* 9 (5): e98401.

Prabhala, R. H., Garewal, H. S., Hicks, M. J., Sampliner, R. E., and Watson, R. R. 1991. "The effects of 13-cis-retinoic acid and beta-carotene on cellular immunity in humans." *Cancer* 67 (6): 1556-1560.

Prior, R. L., and Gu, L. 2005. "Occurrence and biological significance of proanthocyanidins in the American diet." *Phytochemistry* 66 (18): 2264-2280.

Qin, Y., Xia, M., Ma, J., Hao, Y., Liu, J., Mou, H., Cao, L., and Ling, W. 2009. "Anthocyanin supplementation improves serum LDL-and HDL-cholesterol concentrations associated with the inhibition of cholesteryl ester transfer protein in dyslipidemic subjects." *The American Journal of Clinical Nutrition* 90 (3): 485-492.

Rafiq, S., Huma, N., Rakariyatham, K., Hussain, I., Gulzar, N., and Hayat, I. 2018. "Anti-inflammatory and anticancer activities of water-soluble peptide extracts of buffalo and cow milk Cheddar cheeses." *International Journal of Dairy Technology* 71 (2): 432-438.

Ramkisson, S., Dwarka, D., Venter, S., and Mellem, J. J. 2020. "In vitro anticancer and antioxidant potential of Amaranthus cruentus protein and its hydrolysates." *Food Science and Technology* 40: 634-639.

Rao, A., Ray, M., and Rao, L. 2006. "Lycopene." *Advances in Food and Nutrition Research* 51: 99-164.

Rao, A. V., and Agarwal, S. 2000. "Role of antioxidant lycopene in cancer and heart disease." *Journal of the American College of Nutrition* 19 (5): 563-569.

Rao, A. V., and Rao, L. G. 2007. "Carotenoids and human health." *Pharmacological Research* 55 (3): 207-216.

Rashidinejad, A., Birch, E. J., Sun-Waterhouse, D., and Everett, D. W. 2014. "Delivery of green tea catechin and epigallocatechin gallate in liposomes incorporated into low-fat hard cheese." *Food Chemistry* 156: 176-183.

Ray, R. C., and Didier, M. 2014. *Microorganisms and fermentation of traditional foods*. CRC Press.

Read, A., Wright, A., and Abdel-Aal, E. S. M. 2015. "In vitro bioaccessibility and monolayer uptake of lutein from wholegrain baked foods." *Food Chemistry* 174: 263-269.

Ribeiro, A., Caleja, C., Barros, L., Santos-Buelga, C., Barreiro, M. F., and Ferreira, I. C. 2016. "Rosemary extracts in functional foods: Extraction, chemical characterization and incorporation of free and microencapsulated forms in cottage cheese." *Food & Function* 7 (5): 2185-2196.

Ried, K., and Fakler, P. 2011. "Protective effect of lycopene on serum cholesterol and blood pressure: Meta-analyses of intervention trials." *Maturitas* 68 (4): 299-310.

Rizza, S., Muniyappa, R., Iantorno, M., Kim, J.-a., Chen, H., Pullikotil, P., Senese, N., Tesauro, M., Lauro, D., and Cardillo, C. 2011. "Citrus polyphenol hesperidin stimulates production of nitric oxide in endothelial cells while improving endothelial function and reducing inflammatory markers in patients with metabolic syndrome." *The Journal of Clinical Endocrinology & Metabolism* 96 (5): E782-E792.

Roberfroid, M. 2002. "Functional food concept and its application to prebiotics." *Digestive and Liver Disease* 34: S105-S110.

Roberfroid, M. 2007a. "Prebiotics: the concept revisited." *The Journal of Nutrition* 137 (3): 830S-837S.

Roberfroid, M., Gibson, G. R., Hoyles, L., McCartney, A. L., Rastall, R., Rowland, I., Wolvers, D., Watzl, B., Szajewska, H., and Stahl, B. 2010. "Prebiotic effects: metabolic and health benefits." *British Journal of Nutrition* 104 (S2): S1-S63.

Roberfroid, M. B. 2000. "Prebiotics and probiotics: are they functional foods?" *The American Journal of Clinical Nutrition* 71 (6): 1682s-1687s.

Roberfroid, M. B. 2001. "Prebiotics: preferential substrates for specific germs?" *The American Journal of Clinical Nutrition* 73 (2): 406s-409s.

Roberfroid, M. B. 2007b. "Inulin-type fructans: functional food ingredients." *The Journal of Nutrition* 137 (11): 2493S-2502S.

Rodriguez-Amaya, D. B., Kimura, M., Godoy, H. T., and Amaya-Farfan, J. 2008. "Updated Brazilian database on food carotenoids: Factors affecting carotenoid composition." *Journal of Food Composition and Analysis* 21 (6): 445-463.

Rodríguez-Ramiro, I., Ramos, S., Bravo, L., Goya, L., and Martín, M. Á. 2011. "Procyanidin B2 and a cocoa polyphenolic extract inhibit acrylamide-induced apoptosis in human Caco-2 cells by preventing oxidative stress and activation of JNK pathway." *The Journal of Nutritional Biochemistry* 22 (12): 1186-1194.

Roowi, S., Stalmach, A., Mullen, W., Lean, M. E., Edwards, C. A., and Crozier, A. 2009. "Green tea flavan-3-ols: colonic degradation and urinary excretion of catabolites by humans." *Journal of Agricultural and Food Chemistry* 58 (2): 1296-1304.

Rothwell, J. A., Perez-Jimenez, J., Neveu, V., Medina-Remon, A., M'Hiri, N., García-Lobato, P., Manach, C., Knox, C., Eisner, R., and Wishart, D. S. 2013. "Phenol-Explorer 3.0: a major update of the Phenol-Explorer database to incorporate data on the effects of food processing on polyphenol content." *Database* 2013: bat070.

Rudrappa, T., and Bais, H. P. 2008. "Curcumin, a known phenolic from Curcuma longa, attenuates the virulence of Pseudomonas aeruginosa PAO1 in whole plant and animal pathogenicity models." *Journal of Agricultural and Food Chemistry* 56 (6): 1955-1962.

Russo, F., Linsalata, M., and Orlando, A. 2014. "Probiotics against neoplastic transformation of gastric mucosa: Effects on cell proliferation and polyamine metabolism." *World Journal of Gastroenterology: WJG* 20 (37): 13258.

Russo, M., Spagnuolo, C., Tedesco, I., Bilotto, S., and Russo, G. L. 2012. "The flavonoid quercetin in disease prevention and therapy: facts and fancies." *Biochemical Pharmacology* 83 (1): 6-15.

Sahin, K., Tuzcu, M., Gencoglu, H., Dogukan, A., Timurkan, M., Sahin, N., Aslan, A., and Kucuk, O. 2010. "Epigallocatechin-3-gallate activates Nrf2/HO-1 signaling pathway in cisplatin-induced nephrotoxicity in rats." *Life Sciences* 87 (7-8): 240-245.

Sánchez, A., and Vázquez, A. 2017. "Bioactive peptides: A review." *Food Quality and Safety* 1 (1): 29-46.

Santos, A. C., Veiga, F., and Ribeiro, A. J. 2011. "New delivery systems to improve the bioavailability of resveratrol." *Expert Opinion on Drug Delivery* 8 (8): 973-990.

Šaponjac, V. T., Ćetković, G., Čanadanović-Brunet, J., Pajin, B., Djilas, S., Petrović, J., Lončarević, I., Stajčić, S., and Vulić, J. 2016. "Sour cherry pomace extract encapsulated in whey and soy proteins: Incorporation in cookies." *Food Chemistry* 207: 27-33.

Sarma, A. D., Sreelakshmi, Y., and Sharma, R. 1997. "Antioxidant ability of anthocyanins against ascorbic acid oxidation." *Phytochemistry* 45 (4): 671-674.

Satoh, T., Igarashi, M., Yamada, S., Takahashi, N., and Watanabe, K. 2015. "Inhibitory effect of black tea and its combination with acarbose on small intestinal α-glucosidase activity." *Journal of Ethnopharmacology* 161: 147-155.

Sawale, P. D., Patil, G. R., Hussain, S. A., Singh, A. K., and Singh, R. R. B. 2017. "Effect of incorporation of encapsulated and free Arjuna herb on storage stability of chocolate vanilla dairy drink." *Food Bioscience* 19: 142-148.

Scalbert, A., and Williamson, G. 2000. "Dietary intake and bioavailability of polyphenols." *The Journal of Nutrition* 130 (8): 2073S-2085S.

Scarmo, S., Cartmel, B., Lin, H., Leffell, D. J., Welch, E., Bhosale, P., Bernstein, P. S., and Mayne, S. T. 2010. "Significant correlations of dermal total carotenoids and dermal lycopene with their respective plasma levels in healthy adults." *Archives of Biochemistry and Biophysics* 504 (1): 34-39.

Sekhon, B. S., and Jairath, S. 2010. "Prebiotics, probiotics and synbiotics: an overview." *Journal of Pharmaceutical Education & Research* 1 (2).

Setchell, K. 1995. "Discovery and potential clinical importance of mammalian lignans." *Flaxseed in Human Nutrition*: 82-98.

Setchell, K., Borriello, S., Gordon, H., Lawson, A., Harkness, R., Morgan, D., Kirk, D., Adlercreutz, H., Anderson, L., and Axelson, M. 1981. "Lignan formation in man—microbial involvement and possible roles in relation to cancer." *The Lancet* 318 (8236): 4-7.

Shah, N. 2004. "Probiotics and prebiotics." *Agro Food Industry Hi-Tech.*

Shahidi, F. 2009. "Nutraceuticals and functional foods: whole versus processed foods." *Trends in Food Science & Technology* 20 (9): 376-387.

Shahidi F. 2010. "Functional and nutraceutical lipids." In: Smith J, Charter E, editors. *Functional Food Product Development.* UK: John Wiley & Sons Ltd, p. 101–7.

Shahidi, F., and Naczk, M. 1995. *Food Phenolics*. Technomic Pub. Co.

Shimizu, M., and Son, D. O. 2007. "Food-derived peptides and intestinal functions." *Curr Pharm Des* 13 (9): 885-95.

Shu-Yuan, L., Chang, C.-Q., Fu-Ying, M., and Chang-Long, Y. 2009. "Modulating effects of chlorogenic acid on lipids and glucose metabolism and expression of hepatic peroxisome proliferator-activated receptor-α in golden hamsters fed on high fat diet." *Biomedical and Environmental Sciences* 22 (2): 122-129.

Sies, H., Stahl, W., and Sundquist, A. R. 1992. "Antioxidant functions of vitamins." *Annals of the New York Academy of Sciences* 669 (1): 7-20.

Sintara, K., Thong-Ngam, D., Patumraj, S., Klaikeaw, N., and Chatsuwan, T. 2010. "Curcumin suppresses gastric NF-κB activation and macromolecular leakage in Helicobacter pylori-infected rats." *World Journal of Gastroenterology: WJG* 16 (32): 4039.

Skinner, M., and Hunter, D. 2013. *Bioactives in Fruit: Health Benefits and Functional Foods*. John Wiley & Sons.

Song, E.-J., Han, K., Lim, T.-J., Lim, S., Chung, M.-J., Nam, M. H., Kim, H., and Nam, Y.-D. 2020. "Effect of probiotics on obesity-related markers per enterotype: a double-blind, placebo-controlled, randomized clinical trial." *EPMA Journal* 11 (1): 31-51.

Souza, F. P., Balthazar, C. F., Guimarães, J. T., Pimentel, T. C., Esmerino, E. A., Freitas, M. Q., Raices, R. S., Silva, M. C., and Cruz, A. G. 2019. "The addition of xyloligoosaccharide in strawberry-flavored whey beverage." *LWT-Food Science and Technology* 109: 118-122.

Spagnuolo, C., Russo, G. L., Orhan, I. E., Habtemariam, S., Daglia, M., Sureda, A., Nabavi, S. F., Devi, K. P., Loizzo, M. R., and Tundis, R. 2015. "Genistein and cancer: current status, challenges, and future directions." *Advances in Nutrition* 6 (4): 408-419.

Spencer, J. P., Schroeter, H., Kuhnle, G., Srai, S. K. S., Tyrrell, R. M., Ulrich, H., and Catherine, R.-E. 2001. "Epicatechin and its in vivo metabolite, 3′-O-methyl epicatechin, protect human fibroblasts from oxidative-stress-induced cell death involving caspase-3 activation." *Biochemical Journal* 354 (3): 493-500.

Stanton, C., Desmond, C., Fitzgerald, G., and Ross, R. 2003. "Probiotic health benefits-reality or myth?" *Australian Journal of Dairy Technology* 58 (2): 107.

Stodt, U. W., Blauth, N., Niemann, S., Stark, J., Pawar, V., Jayaraman, S., Koek, J., and Engelhardt, U. H. 2014. "Investigation of processes in black tea manufacture through model fermentation (oxidation) experiments." *Journal of Agricultural and Food Chemistry* 62 (31): 7854-7861.

Sueishi, Y., Hori, M., Ishikawa, M., Matsu-Ura, K., Kamogawa, E., Honda, Y., Kita, M., and Ohara, K. 2014. "Scavenging rate constants of hydrophilic antioxidants against multiple reactive oxygen species." *Journal of Clinical Biochemistry and Nutrition* 54 (2): 67-74.

Taha, S., El Abd, M., De Gobba, C., Abdel-Hamid, M., Khalil, E., and Hassan, D. 2017. "Antioxidant and antibacterial activities of bioactive peptides in buffalo's yoghurt fermented with different starter cultures." *Food Science and Biotechnology* 26 (5): 1325-1332.

Tang, Y. Z., and Liu, Z. Q. 2008. "Chemical kinetic behavior of chlorogenic acid in protecting erythrocyte and DNA against radical-induced oxidation." *Journal of Agricultural and Food Chemistry* 56 (22): 11025-11029.

Thompson, L. U., Robb, P., Serraino, M., and Cheung, F. 1991. "Mammalian lignan production from various foods." *Nutrition and Cancer* 16 (1): 43-52.

Tomé-Carneiro, J., Larrosa, M., González-Sarrías, A., A Tomas-Barberan, F., Teresa Garcia-Conesa, M., and Carlos Espin, J. 2013. "Resveratrol and clinical trials: the crossroad from in vitro studies to human evidence." *Current Pharmaceutical Design* 19 (34): 6064-6093.

Trumbo, P., Schlicker, S., Yates, A. A., and Poos, M. 2002. "Dietary reference intakes for energy, carbohydrate, fiber, fat, fatty acids, cholesterol, protein and amino acids." *Journal of the American Dietetic Association* 102 (11): 1621-1630.

Tsao, R. 2010. "Chemistry and biochemistry of dietary polyphenols." *Nutrients* 2 (12): 1231-1246.

Udenigwe, C. C., and Aluko, R. E. 2012. "Food protein-derived bioactive peptides: production, processing, and potential health benefits." *Journal of Food Science* 77 (1): R11-R24.

Valenti, L., Riso, P., Mazzocchi, A., Porrini, M., Fargion, S., and Agostoni, C. 2013. "Dietary anthocyanins as nutritional therapy for nonalcoholic fatty liver disease." *Oxidative Medicine and Cellular Longevity* 2013.

van Breemen, R. B., and Pajkovic, N. 2008. "Multitargeted therapy of cancer by lycopene." *Cancer Letters* 269 (2): 339-351.

Vecchi Brumatti, L., Marcuzzi, A., Tricarico, P. M., Zanin, V., Girardelli, M., and Bianco, A. M. 2014. "Curcumin and inflammatory bowel disease: potential and limits of innovative treatments." *Molecules* 19 (12): 21127-21153.

Verleyen, T., Forcades, M., Verhé, R., Dewettinck, K., Huyghebaert, A., and De Greyt, W. 2002. "Analysis of free and esterified sterols in vegetable oils." *Journal of the American Oil Chemists' Society* 79 (2): 117-122.

Vitaglione, P., Morisco, F., Caporaso, N., and Fogliano, V. 2005. "Dietary antioxidant compounds and liver health." *Critical Reviews in Food Science and Nutrition* 44 (7-8): 575-586.

Vitellio, P., Celano, G., Bonfrate, L., Gobbetti, M., Portincasa, P., and De Angelis, M. 2019. "Effects of Bifidobacterium longum and Lactobacillus rhamnosus on gut microbiota in patients with lactose intolerance and persisting functional gastrointestinal symptoms: A randomised, double-blind, cross-over study." *Nutrients* 11 (4): 886.

Wach, A., Pyrzyńska, K., and Biesaga, M. 2007. "Quercetin content in some food and herbal samples." *Food Chemistry* 100 (2): 699-704.

Wang, H., Liu, Y., Qi, Z., Wang, S., Liu, S., Li, X., Wang, H., and Xia, X. 2013. "An overview on natural polysaccharides with antioxidant properties." *Current Medicinal Chemistry* 20 (23): 2899-2913.

Wang, J., Wu, T., Fang, L., Liu, C., Liu, X., Li, H., Shi, J., Li, M., and Min, W. 2020. "Anti-diabetic effect by walnut (Juglans mandshurica Maxim.)-derived peptide LPLLR through inhibiting α-glucosidase and α-amylase, and alleviating insulin resistance of hepatic HepG2 cells." *Journal of Functional Foods* 69: 103944.

Wang, X., Hasegawa, J., Kitamura, Y., Wang, Z., Matsuda, A., Shinoda, W., Miura, N., and Kimura, K. 2011. "Effects of hesperidin on the progression of hypercholesterolemia and fatty liver induced by high-cholesterol diet in rats." *Journal of Pharmacological Sciences* 117 (3): 129-138.

Wang, Y., Wei, X., Wang, F., Xu, J., Tang, X., and Li, N. 2018. "Structural characterization and antioxidant activity of polysaccharide from ginger." *International Journal of Biological Macromolecules* 111: 862-869.

Wang, Y., Zhao, L., Wang, D., Huo, Y., and Ji, B. 2016. "Anthocyanin-rich extracts from blackberry, wild blueberry, strawberry, and chokeberry: antioxidant activity and inhibitory effect on oleic acid-induced hepatic steatosis in vitro." *Journal of the Science of Food and Agriculture* 96 (7): 2494-2503.

Watson, R. R., and Preedy, V. R. 2010. *Bioactive Foods and Extracts: Cancer Treatment and Prevention*. CRC Press.

Watson, R. R., and Preedy, V. R. 2012. *Bioactive Food as Dietary Interventions for Liver and Gastrointestinal Disease*. Academic Press.

Weihrauch, J. L., and Gardner, J. M. 1978. "Sterol content of foods of plant origin." *Journal of the American Dietetic Association* 73 (1): 39-47.

Whigham, L. D., Watras, A. C., and Schoeller, D. A. 2007. "Efficacy of conjugated linoleic acid for reducing fat mass: a meta-analysis in humans." *American Journal of Clinical Nutrition* 85 (5): 1203-11.

Wildman, R. E. 2016. *Handbook of Nutraceuticals and Functional Foods*. CRC press.

Williamson, G., and Clifford, M. N. 2010. "Colonic metabolites of berry polyphenols: the missing link to biological activity?" *British Journal of Nutrition* 104 (S3): S48-S66.

Xavier, A. A. O., Carvajal-Lérida, I., Garrido-Fernández, J., and Pérez-Gálvez, A. 2018. "In vitro bioaccessibility of lutein from cupcakes fortified with a water-soluble lutein esters formulation." *Journal of Food Composition and Analysis* 68: 60-64.

Xiang, R., Cheng, J., Zhu, M., and Liu, X. 2019. "Effect of mulberry (Morus alba) polyphenols as antioxidant on physiochemical properties, oxidation and bio-safety in Cantonese sausages." *LWT-Food Scienece and Technology* 116: 108504.

Yadav, M., Jain, S., Bhardwaj, A., Nagpal, R., Puniya, M., Tomar, R., Singh, V., Parkash, O., Prasad, G., and Marotta, F. 2009. "Biological and medicinal properties of grapes and their bioactive constituents: an update." *Journal of Medicinal Food* 12 (3): 473-484.

Yaghoubzadeh, Z., Ghadikolaii, F. P., Kaboosi, H., Safari, R., and Fattahi, E. 2020. "Antioxidant activity and anticancer effect of bioactive peptides from rainbow trout (Oncorhynchus mykiss) skin hydrolysate." *International Journal of Peptide Research and Therapeutics* 26 (1): 625-632.

Yang, F., Oz, H. S., Barve, S., De Villiers, W. J., McClain, C. J., and Varilek, G. W. 2001. "The green tea polyphenol (−)-epigallocatechin-3-gallate blocks nuclear factor-κB activation by inhibiting IκB kinase activity in the intestinal epithelial cell line IEC-6." *Molecular Pharmacology* 60 (3): 528-533.

Yazdankhah, S., Hojjati, M., and Azizi, M. H. 2019. "The Antidiabetic potential of black mulberry extract-enriched pasta through inhibition of enzymes and glycemic index." *Plant Foods for Human Nutrition* 74 (1): 149-155.

Yoshino, J., Conte, C., Fontana, L., Mittendorfer, B., Imai, S.I., Schechtman, K. B., Gu, C., Kunz, I., Fanelli, F. R., and Patterson, B. W. 2012. "Resveratrol supplementation does not improve metabolic function in nonobese women with normal glucose tolerance." *Cell Metabolism* 16 (5): 658-664.

Youn, C. K., Park, S. J., Lee, M. Y., Cha, M. J., Kim, O. H., You, H. J., Chang, I. Y., Yoon, S. P., and Jeon, Y. J. 2013. "Silibinin inhibits LPS-induced macrophage activation by blocking p38 MAPK in RAW 264.7 cells." *Biomolecules & Therapeutics* 21 (4): 258-263.

Young, D., and Mine, Y. 2009. "Functional bioactive proteins and peptides in nutrigenomics." *Nutrigenomics and Proteomics in Health and Disease: Food Factors and Gene Interactions*: 129-144.

Yousefi, M., Khorshidian, N., Mortazavian, A. M., and Khosravi-Darani, K. 2019. "Preparation optimization and characterization of chitosan-tripolyphosphate microcapsules for the encapsulation of herbal galactagogue extract." *International Journal of Biological Macromolecules* 140: 920-928.

Yu, M., Chen, M., Gui, J., Huang, S., Liu, Y., Shentu, H., He, J., Fang, Z., Wang, W., and Zhang, Y. 2019. "Preparation of Chlorella vulgaris polysaccharides and their antioxidant activity in vitro and in vivo." *International Journal of Biological Macromolecules* 137: 139-150.

Yuan, Q., Xie, Y., Wang, W., Yan, Y., Ye, H., Jabbar, S., and Zeng, X. 2015. "Extraction optimization, characterization and antioxidant activity in vitro of polysaccharides from mulberry (Morus alba L.) leaves." *Carbohydrate Polymers* 128: 52-62.

Yuca, B., Topçu, İ., Yağcılar-Aydemir, H., Özer, C. O., Kılıç, B., and Başyiğit-Kılıç, G. 2019. "Effects of beta-glucan addition on the physicochemical and microbiological characteristics of fermented sausage." *Journal of Food Science and Technology* 56 (7): 3439-3448.

Zakipour, N., Nasiraie, L. R., Mortazavian, A. M., Khorshidian, N., and Sohrabvandi, S. 2020. "Synbiotic Functional Malt Beverage Supplemented with Inulin and Probiotic Strains." *Current Nutrition & Food Science* 16 (7): 1098-1105.

Zello, G. A. 2006. "Dietary Reference Intakes for the macronutrients and energy: considerations for physical activity." *Applied Physiology, Nutrition, and Metabolism* 31 (1): 74-9.

Zendeboodi, F., Khorshidian, N., Mortazavian, A. M., and da Cruz, A. G. 2020. "Probiotic: conceptualization from a new approach." *Current Opinion in Food Science* 32: 103-123.

Zendeboodi, F., and Mortazavian, A. M. 2021. "Probiotic concept: An overview." In *In Vitro Functionality of Probiotics in Foods*, edited by Mortazavian, A.M., Khorshidian, N. Gomez da Cruz, A. 1-14. New York: Nova Science Publishers.

Zhang, M.-M., Cheng, J. Q., Lu, Y. R., Yi, Z.-H., Yang, P., and Wu, X. T. 2010. "Use of pre-, pro-and synbiotics in patients with acute pancreatitis: a meta-analysis." *World Journal of Gastroenterology: WJG* 16 (31): 3970.

Zhang, T., Wang, F., Xu, H. X., Yi, L., Qin, Y., Chang, H., Mi, M. T., and Zhang, Q.-Y. 2013. "Activation of nuclear factor erythroid 2-related factor 2 and PPARγ plays a role in the genistein-mediated attenuation of oxidative stress-induced endothelial cell injury." *British Journal of Nutrition* 109 (2): 223-235.

Zheng, Y., Lee, S. O., Verbruggen, M. A., Murphy, P. A., and Hendrich, S. 2004. "The apparent absorptions of isoflavone glucosides and aglucons are similar in women and are increased by rapid gut transit time and low fecal isoflavone degradation." *The Journal of Nutrition* 134 (10): 2534-2539.

Zinatizadeh, N., Khalili, F., Fallah, P., Farid, M., Geravand, M., and Yaslianifard, S. 2018. "Potential preventive effect of Lactobacillus acidophilus and Lactobacillus plantarum in patients with polyps or colorectal cancer." *Arquivos de Gastroenterologia* 55 (4): 407-411.

Chapter 4

Effects of Bioactive Compounds on Gastrointestinal Health

Abstract

The prevalence of gastrointestinal diseases has increased in recent years and diet is one of the most significant contributors to the development or prevention of such diseases. Several studies have linked bioactive food compounds as helpful tools to decrease these health problems. Bioactive compounds such as probiotics, prebiotics, vitamins, minerals, bioactive lipids, carbohydrates, peptides, proteins, and phytochemicals can be used as functional food ingredients and can exert various health benefits. The human gut microbiota is the target site of action for bioactive food components. This chapter reviews the potential of different food bioactive compounds in the mitigation of gastrointestinal tract diseases.

Keywords: bioactive compound, gastrointestinal health, health benefits, inflammation, nutraceutical

1. Introduction

Over the past decades, the importance of nutritional bioactive compounds in maintaining health has been highlighted. Bioactive compounds are essential or non-essential compounds found in nature or are formed during the processing of food or medicinal plants that exert various biological activities and have beneficial effects on health (Biesalski et al. 2009). Bioactive compounds are derived from different sources but majorly are obtained from plants and include probiotics, prebiotics, peptides, bioactive lipids, carbohydrates, vitamins, minerals and various phytochemicals (Perez-Gregorio and Simal-Gandara 2017). These bioactive compounds are usually served in functional foods and beverages. Consumption of functional foods is associated with advantageous physiological effects such as antioxidant, anti-inflammatory, anti-aging, antimicrobial, anti-obesity, anticancer, neuroprotective, cardioprotective and lipid-lowering activities (Adefegha 2018; Gul, Singh, and Jabeen 2016).

Among the various organs that may be affected by food components, the gastrointestinal (GI) tract has an important position because it is an obligatory passage for ingested foods (Roberfroid 1996). GI is sterile at birth and the microbiota develops after birth with the colonization rate based on the factors including the mode of birth, infant nutrition, using antibiotics, diet and age (Cencic and Chingwaru 2010). Since the human gut has the role of breaking down foods and facilitating the absorption of nutrients, its malfunction negatively affects human health (Cencic and Chingwaru 2010).

The incidence of gastrointestinal diseases such as irritable bowel syndrome (IBS), inflammatory bowel disease (IBD), Crohn's disease (CD), ulcerative colitis (UC) and celiac disease have increased in recent years and besides the role of genetics, the development of these disorders is also influenced by diet, lifestyle and the gut microbiota (Elmaliklis et al. 2019). Recent studies have proved that some categories of functional foods such as probiotics, prebiotics and phytochemicals have a possible positive effect on the gastrointestinal system. This chapter reviews the beneficial effects of various bioactive compounds on gastrointestinal diseases.

2. Therapeutic Effects of Functional Bioactive Compounds on GI Disorders

2.1. Esophagus and Stomach

This section discusses the effects of various bioactive compounds in different diseases of the esophagus and stomachs. Undoubtedly one of the most common diseases in this body part, is peptic ulcer.

2.1.1. Peptic Ulcer

Since peptic ulcer has no definite treatment, various herbal compounds have been studied for this, especially polyphenols. The complex process of gastric ulcer healing is a combination of wound healing and re-epithelialization, which involves intervening multiple factors, especially growth factor and prostaglandins, as key contributors in tissue restoration and angiogenesis (Hosein et al. 2014; Banerjee et al. 2008).

Many of the beneficial effects of traditional and complementary medicines in the treatment of peptic ulcers can be attributed to the presence of polyphenols. Dietary polyphenols, among a wide range of natural molecules, can play a crucial role in treating gastric and duodenal ulcers through numerous biological mechanisms (Rahimi, Abbasabadi, and Abdollahi 2013; Hosein et al. 2014).

One of the effective polyphenols in peptic ulcers healing is curcumin. The anti-ulcerative effects of curcumin are comparable to ranitidine and omeprazole (Chattopadhyay et al. 2006). Intraperitoneal administration of curcumin could relieve the severity of duodenal ulcers induced by indomethacin. Curcumin, as an antioxidant, can have protective effects on the gastrointestinal tract by increasing the level of antioxidant enzymes such as catalase (CAT) and glutathione peroxidase (GPx). On the other hand, curcumin reduces the production of superoxide radicals and reactive oxygen species (ROS) by suppressing the activity of xanthine oxidase and myeloperoxidase (MPO). Thus, curcumin, via antioxidant effects, suppresses stress-related cell apoptosis in the gastrointestinal tract (Sivalingam et al. 2007). Curcumin has been shown to alleviate peptic ulcer and this effect was attributed to the role of curcumin in reducing oxidative damage, preventing the depletion of GPx reserves, scavenging ROS, suppressing lipid and protein peroxidation and eventually protecting gastric mucosa by suppressing raised gastric acid secretion (Swarnakar et al. 2005). Another prominent therapeutic effect of curcumin is its ability to eradicate *Helicobacter pylori* and its gastric damage, as evidenced by in vivo and animal studies (Foryst-Ludwig et al. 2004; De et al. 2009; Zaidi et al. 2009; Sintara et al. 2010; Santos et al. 2015).

In a recent trial, it has been revealed that the combination of curcumin with standard *H. pylori*-positive peptic ulcer treatment can significantly ameliorate dyspepsia symptoms while did not affect the infection eradication (Khonche et al. 2016). Curcumin can suppress the production of IL-8 through the inactivation of NF-κB, a key factor in the pathogenesis of *H. pylori* infection (Münzenmaier et al. 1997). De et al. (2009) demonstrated the efficacy of curcumin in eradicating *H. pylori* infection and curing peptic ulcer by designing an experimental study. Curcumin also reduces gastric acid production by blocking H_2 receptors that followed by a reduction of histamine level (Kim, Kim, et al. 2005).

Another polyphenol with positive anti-ulcer effects is Epigallocatechin gallate (EGCG), mainly extracted from green tea. The ability of EGCG to reduce the production of key inflammatory mediators, which play an essential role in ulcer formation, is the major mechanism for its action. EGCG has a remarkable therapeutic benefit for gastric ulcer via decreasing nitric oxide (NO) level as a stimulant for oxidative epithelium damage and modifying the inducible nitric oxide synthase (iNOS)/endothelial nitric oxide synthase (eNOS) ratio at the gastric tissue (Adhikary et al. 2011). In addition to alleviating oxidative stress, which possesses a substantial role in initiating and exacerbating peptic ulcer, EGCG also suppresses lipid and protein oxidation in ulcerated gastric tissue (Farzaei, Abdollahi, and Rahimi 2015).

Quercetin, a polyphenolic substance from the subgroup of flavonols, and its derivatives are under experimental investigations to evaluate therapeutic efficacy for peptic ulcers (Manach et al. 2005; Conquer et al. 1998; Kahraman et al. 2003). Pretreatment with oral administration of quercetin can mitigate the toxicity of gastric injury-causing agents, which disturb the mucosal and interstitial matrix. Among the involved mechanisms, suppression of free radicals, reduction of histamine levels as a stimulant for gastric secretion, upgrading of the level of antioxidant enzymes such as CAT and superoxide dismutase (SOD) and inhibition of the activity of pro-oxidant enzymes such as MPO can be mentioned (Kahraman et al. 2003; Tandon et al. 2004).

Resveratrol, the most famous stilbenoid (polyphenol), has potential in peptic ulcers treatment owing to anti-oxidative stress and anti-inflammatory activities (Tomé-Carneiro et al. 2013; Panaro et al. 2012). The key mechanism of resveratrol's protective effect in peptic ulcers appears to be due to its potential to suppress the production of major inflammatory mediators. *In vitro* studies have exposed inhibitory effects of resveratrol on the expression and activation of nuclear factors (NF-κB) as well as intracellular transcriptional enzymes like mitogen-activated protein kinase (MAPKs) (Panaro et al. 2012). In an animal model of acetic acid-induced gastric ulcer, it has been shown that resveratrol possesses protective and therapeutic effects through inhibiting the activity of MPO and malondialdehyde (MDA), which respectively indicate a reduction of leukocyte infiltration and lipids peroxidation. Resveratrol also improved collagen content and depleted glutathione (GSH) restoration (Solmaz et al. 2009).

Although suppression of tumor necrosis factor α (TNF-α), a key pro-inflammatory factor in peptic ulcers, is the central mechanism of resveratrol therapeutic effects and the antibacterial effects of resveratrol against *H. pylori* infection are also well recognized. The inhibitory effects of resveratrol on both urease and virulence activities of *H. pylori* are dose-dependent (Mahady, Pendland, and Chadwick 2003; Paulo et al. 2011). The protective effects of hesperidin have been shown in peptic ulcers; although studies in this area are limited, this remains an area of further research. Hesperidin, the major flavonoid in citrus, exerts antioxidant and anti-inflammatory effects by reducing the expression of cyclooxygenase-2 (COX-2), decreasing lipid peroxidation and inhibiting TNF-α production, which entirely results in avoidance of ulcers aggravation (Selmi et al. 2017).

Regarding the role of *H. pylori* in the pathogenesis of peptic ulcers and probiotic effectiveness in eradicating the infection, there is a special interest in probiotics in treating peptic ulcers (Elitsur and Yahav 2005). The efficacy of probiotics in various gastrointestinal diseases and antibiotic-resistant infections has been established (Suerbaum and Michetti 2002). Although some studies have failed to demonstrate the ability of probiotics to eradicate the infection, they have suggested that probiotics can reduce bacterial load and infection (Sgouras et al. 2005). In some human studies, it has been indicated that probiotics can improve the effectiveness of treatment against *H. pylori* elimination and likely improve symptoms of dyspepsia (Vasiljevic and Shah 2008).

Probiotics, as microorganisms, exert a significant role in balancing the micro-ecological environment of the stomach (World Health Organization and Food & Agriculture Organization 2002). The adjuvant therapy of probiotics with standard triple therapy for peptic ulcers improves treatment efficacy and minimizes therapy-related adverse events. Among these benefits are the following: enhancement of the eradication rate of *H. pylori* infection, alleviating clinical symptoms, and the increased recovery rate of patients. Therefore, it can be said that probiotic administration is a worth promotion (Ma et al. 2015; Lu et al. 2016).

Since the balance of gut microflora in various gastrointestinal diseases, including gastric pathologies such as infection with *H. pylori* changes, dietary modulation with adding probiotics can be an effective treatment to restore the normal gastric microbiota (Andersson et al. 2008). The major strains of probiotic bacteria used include *Lactobacillus* and *Bifidobacterium*, and sometimes yeasts such as *Saccharomyces boulardii* (Saxelin et al. 2005).

The major mechanisms by which probiotics thereby repair gastric mucosa in the *H. pylori*- associated peptic ulcer are listed below:

- Probiotics can neutralize the reaction of *H. pylori*-activated cytokines. As it has been shown, *Lactobacillus acidophilus* can deactivate Th1 response to *H. pylori* infection (Yang et al. 2012).
- *Lactobacillus* bacteria by lactic acid secretion attenuate the *H. pylori*-induced hypochlorhydria. On the one hand, this effect reinforces the gastric defense mechanism by luminal pH reduction; on the other

hand, lactic acid secretion also has a direct bactericidal effect (De Keersmaecker et al. 2006). With the exact mechanism, *L. acidophilus* inhibits the urease activity of *H. pylori* (Coconnier et al. 1998).

- Probiotics, particularly *Lactobacillus* strains, compete with pathogenic bacteria to bind to host surface receptors and prevent the adherence of pathogens to epithelial cells. This effect is probably mediated by increasing mucin production (Mack et al. 1999).
- Probiotics have both direct and indirect bactericidal and bacteriostatic effects. For instance, *Lactobacillus reuteri* with a secretion of a metabolite called reuterin against gram-positive and gram-negative bacteria, fungi and yeasts can destroy those (Spinler et al. 2008).

In addition to commercially available products, traditional fermented foods and beverages such as yogurt and kefir are natural sources of probiotics that have anti-*H. pylori* activity (Nair et al. 2016).

Dietary fiber, by acting as a buffer, reduces the concentration of bile acids in the stomach. Besides reducing transit time, it results in less abdominal bloating, discomfort, and pain in the gastrointestinal tract. Therefore, according to the World Health Organization, a high-fiber diet (20 to 30 g/day) is recommended for individuals with peptic ulcer (Marotta and Floch 1991).

For years, in comparing the highest and lowest quintiles of dietary fiber intake, the inverse relationship between dietary fiber intake and the risk of duodenal ulcers has been identified in US men. Also, it has been demonstrated that a diet rich in soluble fiber is strongly associated with reducing the risk of duodenal ulcers (Aldoori et al. 1997). Later, the protective effect of dietary fiber on duodenal and gastrointestinal ulcers has been proven in further studies (Anderson et al. 2009; Ryan-Harshman and Aldoori 2004).

2.1.2. Gastroesophageal Reflux Disease (GERD)

Although the underlying mechanism of gastroesophageal reflux has not yet been completely determined, there is evidence of the role of oxidative stress and lipid peroxidation in the pathogenesis of the disease (Kawahara et al. 2007; Singh et al. 2012). In this respect, various studies have pointed to the role of ROS in triggering esophageal lesions, which, by oxidation of membrane unsaturated fatty acids, leads to lipid peroxidation (Singh et al. 2012; Aiyer et al. 2011). Therefore, reinforcement of antioxidant defense systems, including antioxidant enzymes, can possess a beneficial and protective role in preventing toxic effects of ROS (Singh et al. 2012). These reports provided new insight into applying natural products that regulate oxidation and inflammatory state in the treatment of GERD, and led toward studies aimed at the investigation (Kang and Lee 2014).

Phenolic compounds-rich plants are mainly used in the production of new drugs due to their antioxidant ability and they have promising results in the treatment of GERD (Kawahara et al. 2007; Mahattanadul et al. 2011). Chlorogenic acid (CGA), as a phenolic substance, can suppress the drug-induced gastric mucosal lesion (Shimoyama et al. 2013; Stalmach et al. 2010; Farah et al. 2008).

Chlorogenic acid can attenuate the severity of esophageal lesions through the down-regulation of inflammatory response. Histopathological studies have attributed these beneficial effects to the reduction of lipid peroxidation, serum TNF-α level, and the expression of COX2 and NOS enzymes and to increase the reduced glutathione/oxidized glutathione ratio. Therefore, the administration of CGA, due to its antioxidant capacity, as an alternative treatment and/or standard anti-secretory therapies can reduce the severity of GERD and reverse the progression of the disease (Kang and Lee 2014).

Owing to antioxidant effects, quercetin can attenuate the severity of esophagitis triggered by reflux. It has also been revealed to reduce the level of collagen in the damaged esophageal tissue of omeprazole. The effects of quercetin and α-tocopherol on reducing the elevated level of collagen in damaged esophageal tissue are comparable to omeprazole. These results indicate that the therapeutic use of quercetin is validated in gastroesophageal reflux disease (Rao and Vijayakumar 2008). The finding of a clinical trial study indicated that β-carotene therapy could successfully reduce GERD symptoms. Beta-carotene also decreased the requirement of anti-reflux medication in 50% of patients (Dutta et al. 2012).

Long-term gastric acid and bile reflux causes inflammation and mucosal damage in the esophagus, which ultimately can lead to esophagitis and Barrett's esophagus. The gastroesophageal reflux impairs the mucosal barrier and causes exposure to epithelial cells to orally swallowed microbes, regurgitated gastric acid and bile acid contents (Yang, Francois, and Pei 2012).

Microbiome composition is assumed to be altered in GERD, due to the widespread use of antibiotics. Prolonged microbial imbalance can be carcinogenic and its association with esophageal adenocarcinoma has been reported (Pei et al. 2004; Pei and Yang 2005). In the pathogenesis of esophageal diseases, change in microbial composition is more important rather than an absolute load of bacteria (Yang et al. 2014). The forward flow of bowel contents decreases the opportunity of GERD by reducing stomach pressure. Esophageal peristalsis is influenced by lipopolysaccharide (LPS) secreted from the bacteria, which reduces the base tone (Fan et al. 2001). In the animal study, it has been shown that treatment with LPS prevents proper gastric emptying by lowering the normal gastrointestinal motility and causes the stomach to remain complete full between meals in LPS-treated mice (Calatayud et al. 2002).

Co-administration of probiotic strain, *Lactobacillus gasseri*, with proton pump inhibitors in patients with reflux improved upper GI tract-related symptoms that were resistant to medication therapy alone. Unfortunately, there was no control group in the present study, so the interpretation of the results should be done with caution (Igarashi et al. 2014). The protective and positive benefits of fiber consumption on multiple gastrointestinal diseases, including GERD, peptic ulcer, biliary disease, diverticulitis, and colorectal cancer, have been proven (Trowell 1975).

The available evidence from clinical trials and epidemiological studies strongly supports the association of high fiber intakes with lower prevalence of duodenal ulcer disease (Ryan-Harshman and Aldoori 2004). However, the population-based case-control study has not been able to demonstrate a relation between fiber and total carbohydrate intake with the risk of reflux-induced esophagitis (Mulholland et al. 2009). The most negligible protective effects of fiber consumption can attenuate the patient's clinical symptoms (El-Serag, Satia, and Rabeneck 2005). Furthermore, the fiber in particular soluble fiber, reduces gastric acid production (El-Serag, Satia, and Rabeneck 2005). Owing to the capacity of fiber to absorb dietary carcinogens, reverse association of fiber intake from fruits and vegetables with the risk of GERD and Barrett's esophagus can be supposable (Kubo et al. 2009; Morris Brown et al. 1995).

2.2. Intestine and Colon

The small and large intestine disorders include a large part of the gastrointestinal disorders described entirely in the second chapter. Here the effects of active compounds on treating these diseases are discussed.

2.2.1. Diarrhea

In the old days, there have been numerous herbal compounds to treat diarrhea. Recent investigations on the efficacy of plants in disease management and medicine have exposed that various phytochemicals can affect the function and motility of the intestine. Also, some of them have antibacterial activity

(Njume and Goduka 2012; Bukhari et al. 2013; Velázquez et al. 2012; Rajan et al. 2012; Abbassi and Hani 2012; Knipping, Garssen, and van't Land 2012; Mariita et al. 2011). It has also been revealed that the consumption of bioactive polyphenols can reduce the frequency of gut movements and restore fecal consistency rapidly to normal (Dover, Patel, and Park 2015). Traditional use of the stem of *Lantana camara* to treat diarrhea confirms these scientific findings. The chemical examination of plants shows that phytochemical extracts, including flavonoids, alkaloids, tannins, and phytosterols play a principal role in anti-diarrhea activity (Tadesse et al. 2017). Many cases of diarrhea are due to infection. Recent evidence suggests that hesperidin and ellagic acid, due to their immunomodulating and anti-microbial activities, can act as a diarrhea remedy (Abuelsaad et al. 2013).

Flavonoids have potential in treating diarrhea, which can be due to their inhibitory effects on the secretion of prostaglandins and the improvement of the absorption of electrolytes (Awad, Toczek, and Fink 2004; Hämäläinen et al. 2011). One of the most effective flavonoids from this perspective is tannins. Tannins with an effect on GI mucosa inhibit calcium-dependent chloride channels (CaCC) and cystic fibrosis transmembrane regulator (CFTR), which can reduce the secretion of fluids (Ashok and Upadhyaya 2012; Wongsamitkul et al. 2010). Although the main biological effects of flavonoids are attributed to the free radical-scavenging ability and their antioxidant properties, various investigations have stated anti-infective, anti-inflammatory and anti-allergic activities. That could be a justification for the protective role of flavonoids in heart disease, cancer, and multiple pathologies. However, it should be noted that flavonoids at very high doses can exert pro-oxidant effects (Martínez-Flórez et al. 2002; Matsuda et al. 2003).

Curcumin, a naturally occurring plant phenolic compound, has shown beneficial synergistic effects with antibiotics against bacteria-induced diarrhea (Sasidharan et al. 2014). Most of the beneficial effects of curcumin may be through improving the function of the gut epithelial barrier and reducing its permeability. Substantial weight gain, bloating and abdominal pain reduction in patients with HIV-associated diarrhea confirm these findings (Conteas et al. 2009). Therefore, curcumin can be used as a combination therapy for diarrhea.

In recent years, the interaction between host and microbiota and the consequences of the microbiota imbalance to health have been widely considered (Shanahan 2002; Isolauri, Salminen, and Ouwehand 2004). There is strong evidence of the efficacy of probiotics as an effective remedy for gastroenteritis and fluid therapy (Guandalini 2011). Administrating probiotics in combination with rehydration can shorten the course of diarrhea, particularly infection-associated diarrhea in children (Saavedra et al. 1994). The main mechanism of probiotics in controlling diarrhea is the ability to inhibit the growth of pathogens and reduce their adhesion to the intestinal epithelium. However, the mechanism of probiotic action in treating various types of diarrhea with different etiologies has not yet been fully elucidated. In recent clinical trials, several mechanisms have been proposed for probiotics, including production of antimicrobials such as organic acids, competition with pathogens in nutrient consumption, competition with pathogens for adhesion sites, strengthening the gut epithelial barrier, immunomodulatory effects by stimulating the synthesis of immunoglobulin A (IgA) and stimulating Th1 response by producing cytokines like Il-12, Il-2 and particularly IFN-α. In conclusion, probiotics, by inhibiting the formation of toxins, exert their cytoprotective effects (Levy 2000; Isolauri et al. 2001). Concerning this, probiotics especially administered at the onset of infectious diarrhea may be effective in limiting the course of the illness (Allen et al. 2003; Guandalini 2011; Floch et al. 2011).

Taking antibiotics with an alteration in intestinal flora can increase gut motility and cause abdominal cramps. Administration of probiotics in this case by restoring the balanced gut microflora prevents diarrhea or reduces its duration and severity (Cimperman et al. 2011). *Clostridium difficile* colitis is considered one of the most common types of antibiotic-induced diarrhea, a scourge of daycare centers, hospitals and nursing homes (Kelly and LaMont 2008). Numerous studies have pointed to the influential role of specific strains of probiotics such as *S. boulardii*, *L. acidophilus* and *Bifidobacterium* in preventing and treating antibiotic-associated diarrhea (D'souza et al. 2002; McFarland 2006; Floch et al. 2011; Hempel et al. 2012). A 2011 Cochrane review of 16 studies concluded that the use of *Lactobacillus rhamnosus* GG and *S. boulardii* is an appropriate treatment for diarrhea, but further large, well-designed, randomized control trials are needed to confirm these findings (Johnston et al. 2007). However, despite the well-established benefits of prebiotics, such as inulin and fructooligosaccharides to preserve the balance and health of the intestinal microflora, there is not enough evidence for the prescription of prebiotics, and only a few limited animal studies have been conducted in this area (de Vrese and Marteau 2007).

Except for probiotics, the positive effects of dietary fiber on diarrhea have been studied copiously. Based on the recent clinical trial, co-administration of fiber and probiotics by improving intestinal movements and reducing digestive disorders have extra benefits in reducing the incidence of diarrhea, especially in patients on enteral nutrition (Zhao et al. 2017). In another study, supplementation of tube-fed full-resuscitated and mechanically ventilated septic patients with soluble fiber reduced the incidence of diarrhea (Spapen et al. 2001). Similar promising results were found in infants over the age of 6 months that receiving fiber-supplemented soy formula could reduce the duration and symptoms of acute diarrhea (Vanderhoof et al. 1997). Administration of soluble dietary fiber in elderly patients with diarrhea and intestinal atrophy could ameliorate their symptoms and improve the nutritional status through enhancing bowel movements, normalizing gut microflora and repairing the intestinal mucosa (Nakao et al. 2002).

With scientific findings, the traditional use of flaxseed as a diarrhea remedy is now justifiable. As a functional food, Flaxseed has a wide range of health benefits because of its medicinally active constituents. About 40–50% of flaxseed is made up of omega-3 rich oil, 23-34% from protein and 5% from soluble fiber, and contains lignin precursors (Tarpila, Wennberg, and Tarpila 2005; Duke et al. 2002; Joerg et al. 2000).

Animal studies have revealed promising results of the positive effects of flaxseed on both infectious and non-infectious diarrhea. Besides the ability to cope with pathogens, flaxseed possesses predominant antispasmodic activity through blocking calcium channels. This, coupled with the effect of flaxseed on consistency and frequency of stool and by controlling gastrointestinal movements and secretions can be an effective treatment for diarrhea (Palla et al. 2015).

2.2.2. Constipation

Constipation is one of the gastrointestinal disorders that, due to the chronic nature of the disease, patients experience a variety of chemical and herbal treatments, each with various advantages and disadvantages.

Flaxseed in the form of whole grains or extracted oil and mucilage, can be used for different purposes (Kaithwas et al. 2011). Flaxseed can be medically recommended in the treatment of both diarrhea and constipation. This tremendous bilateral effect on the regulation of gut motility is possibly due to the coexisting of soluble and insoluble fibers. Although clinical trials have shown that flaxseed is laxative as a whole, bulking properties are mainly attributed to the insoluble fiber of the grain (Duke et al. 2002; Xu et al. 2012; Tarpila, Wennberg, and Tarpila 2005). Other than fiber, flaxseed oil can act as a lubricant laxative agent that enhances the effect of the seed in alleviating constipation (Xu et al. 2012).

The association of constipation with intestinal microbiota dysbiosis has been confirmed in experimental studies (Attaluri et al. 2010). This imbalance manifests itself as a reduction in the abundance of good strains, including bifidobacteria, *Lactobacillus* and *Bacteroides*, and an increase in the number of pathogens (Khalif et al. 2005; Zhao and Yu 2016; Zhu et al. 2014). From this perspective, restoring normal intestinal microflora can be an effective way of preventing and treating constipation. Supplementation with probiotics or consumption of probiotic-enriched food products such as bio-yogurt can increase counts of *Lactobacillus* and *Bifidobacterium* in the intestines.

Increasing the number of beneficial bacteria can help manage constipation in several ways. First, producing short-chain fatty acids (SCFAs) improves the motor activity of the gastrointestinal tract by stimulating the contraction of the muscles of the intestinal wall and normalizing gut motility. Gut sensation also alters with intestinal microbiota modification by probiotics. Second, SCFAs, in addition to being absorbed from intestinal mucosa and having systemic effects, are also used by intestinal microbiota for biomass formation, which will increase fecal mass and accelerate the intestinal passage. Finally, it seems that one of the imperative factors that lead to feces evacuation facilitation following probiotic supplementation is the high water content (~80%) in the biomass and the ability of these bacteria to produce gas. Furthermore, probiotics with an effect on the secretion and absorption of water and electrolytes, the reduction of the intestinal pH and the production of specific metabolites can regulate the intraluminal environment (Zhao and Yu 2016; Zhu et al. 2014; Ouwehand et al. 2002; Koebnick et al. 2003; Salminen and Salminen 1997). However, the effect of probiotics on intestinal microbiota in constipation is still not fully understood. Except for gastrointestinal microbiota and short-chain fatty acids, regulation of intestinal motility is also influenced by other factors, including luminal factors and hormones; for instance, bile acids as physiological laxatives can modulate water and electrolytes luminal transport (Keely et al. 2007; Snape, Shiff, and Cohen 1980).

The benefits of prebiotics in enhancing the efficacy of constipation treatment should not be overlooked. Prebiotics increase the osmolality of feces through binding water. Although it has been shown that some prebiotics, such as lactulose and oligosaccharides, have potent laxative effects, but more studies in this area are needed.

2.2.3. Irritable Bowel Syndrome (IBS)

Up to now, an effective and safe medication for irritable bowel syndrome is not identified; on the other hand, the relationship between intestinal microflora and IBS has been proven. Therefore, certain probiotic strains seem to be practical therapeutic choice for patients suffering from this syndrome.

Intestinal microbiota in IBS patients compared to healthy subjects indicates the role of intestinal microflora imbalance in IBS pathophysiology (Kassinen et al. 2007; Fukudo et al. 2015). Hence, interventions aimed to restore the balance of intestinal flora in IBS patients, mainly probiotics, have been increasingly considered in recent years (Distrutti et al. 2016).

In addition to altering the intestinal microflora, probiotics can exert beneficial effects on the treatment of IBS through mechanisms that target gut barrier integrity, visceral hypersensitivity, gut movements, and intestinal immune function enhancement (Simrén et al. 2012). Assessment of RCTs in systematic review and meta-analysis studies suggests that probiotics can effectively treat IBS (Ford, Quigley, et al. 2014; Didari et al. 2015; Zhang, Li, et al. 2016). However, RCTs have heterogeneity in terms of the study design, strains of examined probiotics and the approach of strains are administered to patients. Therefore, comparing these studies with meta-analysis is somewhat difficult. Based on the results of the studies, it seems that each probiotic strain is effective in alleviating specific symptoms. For example, some of which appear to mainly reduce abdominal cramp and flatulence (Nobaek et al. 2000; Kim, Vazquez Roque, et al. 2005), Some are effective in intestinal hyperactivity (Dolin 2009), whereas certain strains have positive effects on improving systemic symptoms of the patient (Enck et al. 2008; Guglielmetti et al. 2011).

Despite extensive investigations on the effects of probiotics in IBS, the optimal therapeutic approach remains unclear in terms of effective probiotic type (single or mixtures of species), the perfect dosage and duration of treatment (Lacy, Chey, and Lembo 2015; Harris and Baffy 2017). Given the lack of consistent data, the American College of Gastroenterology strongly recommends the administration of probiotics to patients with IBS (Ford, Moayyedi, et al. 2014). Therefore, there is no doubt that there is a need for high-quality RCT studies to provide further evidence about the effectiveness of probiotics as a therapeutic option in managing IBS symptoms. On the whole, based on the mentioned statements from literature reviews and following dietary guidelines, it can be concluded that although probiotics appear to exert several health benefits in IBS, it is still not well-proven that for each person which probiotic strains, at what dose and how long should be prescribed (Lacy, Chey, and Lembo 2015). Patients are also advised to choose a certain probiotic product at a time and monitor the effects for at least four weeks. Then, if they obtain positive feedback from it, they can continue to use, and if probiotic does not work, its type should be changed (McKenzie et al. 2016).

A logical approach to enhancing favorable intestinal bacteria growth, such as bifidobacteria, is using prebiotics. While prebiotics can partially modify dysbiosis in IBS patients, a growing body of evidence has shown a limited effect since a high dose of some prebiotics, such as oligofructose and inulin, may exacerbate the symptoms of patients due to colonic gas-producing via bacterial fermentation. Due to controversy regarding the therapeutic potential of prebiotics in IBS, the American College of Gastroenterology has announced that evidence for prebiotics in patients suffering from IBS is insufficient (Ford, Moayyedi, et al. 2014). Therefore, additional studies are needed to determine the type, dosage and duration of optimum use of probiotics.

Dietary fiber includes non-digestible polysaccharides and lignin, which may possess many health benefits (Chutkan et al. 2012). The treatment of many gastrointestinal disorders requires an alteration in dietary fiber, which is mainly an incremental change (Ramkumar and Rao 2005; Fernandez-Banares et al. 1999). Irritable bowel syndrome is not the exception and up to now, a considerable number of studies have been published on the role of dietary fiber in the IBS (Philpott et al. 2015; Nagarajan et al. 2015; Rao, Yu, and Fedewa 2015; Ford, Moayyedi, et al. 2014; Moayyedi et al. 2014). The role of fiber in IBS is still subject to ongoing debate owing to conflicting findings and a lack of high-quality investigations. Various reviews have highlighted a deficient intake of fiber in patients with IBS (Ford et al. 2008). Accordingly, increased fiber intake is recommended for all patients with IBS, particularly IBS-C (El-Salhy 2012; Ford, Moayyedi, et al. 2014). However, a meta-analysis of 12 studies failed to demonstrate the beneficial effects of dietary fiber (either soluble or insoluble) on improving IBS symptoms over placebo or a low-fiber diet (Ford et al. 2008). The inconsistencies in the finding of recent studies indicate that it is necessary to consider the classification of IBS, i.e., IBS with diarrhea (IBS-D), IBS with constipation (IBS-C) and mixed IBS (IBS-M), in dietary recommendations in addition to the type of fiber consumed. In most studies, it has been shown that consuming soluble fiber mitigates overall IBS symptoms while increasing the intake of insoluble fibers does not have such an effect (Francis and Whorwell 1994; Bijkerk et al. 2009).

To better understand the mentioned issue, it is required to consider the physical and chemical characteristics of different types of fibers. Accordingly, it is possible to decide which type of fiber is appropriate for each IBS subclass. Classification of fiber into two groups of soluble and insoluble in terms of fiber efficacy in IBS with the recommendation to respectively increase and decrease their intake does not seem reasonable (Moayyedi et al. 2014; McKenzie et al. 2016). The classification of fibers based on viscosity, gel formation and fermentability is preferable to solubility in water. The first two properties are clear, but in explaining the third one, readily fermented fibers rapidly cause fluctuating, while poorly fermented fibers lead less gas production (McRorie Jr 2015). Given that, fibers are classified into four groups as follows:

1. Insoluble, non-viscous, poorly fermented such as wheat bran.
2. Water-soluble, non-viscous, readily fermented such as inulin.
3. Soluble viscous/gel-forming, readily fermented such as β-glucan.
4. Soluble viscous/gel-forming, non-fermented such as psyllium.

It should be noted that certain foods contain a combination of different fiber types (Capili, Anastasi, and Chang 2016).

Soluble viscous fibers with a low rate of fermentation, such as psyllium, can be recommended to patients with all IBS subtypes (Chutkan et al. 2012; Bijkerk et al. 2009; Moayyedi et al. 2014). In other words, psyllium would be applicable to treat both constipation and diarrhea in IBS patients (Capili, Anastasi, and Chang 2016). A recent systematic review and meta-analysis including 14 RCTs confirmed this statement. They concluded that soluble fiber could benefit IBS patients, while insoluble fibers such as bran have no favorable effect or may even exacerbate symptoms (Moayyedi et al. 2014). Therefore, the general recommendation for increasing dietary fiber for all of these patients is not rational (Ford et al. 2008).

Association of gastrointestinal symptoms such as bloating, abdominal cramps and change in the bowel habits with the onset of fiber supplementation has been well established, particularly in patients with IBS (Brandt et al. 2005; Moayyedi et al. 2014; El-Salhy and Gundersen 2015). Therefore, increasing dietary fiber intake should be gradual so that the weakly increase should not exceed 5 grams (Agréus et al. 1995). Another rich source of fiber is flaxseed, which can be recommended to treat constipation by adding it to a diet. A recent study conducted in IBS-C patients revealed that ground flaxseed (6-24 g/d) gradually supplementation over three months may effectively reduce abdominal pain and distension (McKenzie et al. 2016).

Fiber can exert a mechanical stimulation of the intestinal mucosa and increase the bulk of stool, which may change bowel habits. The bacterial fermentation of the fiber in the colon with the production of SCFA reduces the pH of the lumen environment, and on the other hand, increases the gas production and the pressure of the lumen, which both affect the enteric nervous system and consequently affect the digestive secretion and gut motility. Along with local effects, fibers also have systemic benefits, including lowering cholesterol levels, improving glycemic regulation, and weight control (Chutkan et al. 2012; Eswaran, Muir, and Chey 2013; McRorie et al. 2000). In general, based on literature reviews and following dietary guidelines, it can be concluded that fiber intake should be increased gradually to reach the target of 20-30 g per day. However, the optimal fiber intake is still unclear (Eswaran, Tack, and Chey 2011).

Psychological distress and hormonal fluctuations are associated with the functional gut disorder, including IBS, characterized by symptoms such as visceral hypersensitivity, intestinal hyperpermeability and abdominal cramp (Longstreth 2005; Thompson et al. 1999; Ritchie 1973; Spiller et al. 2000). On the other hand, it has been discovered that increased intestinal permeability is related to the severity of IBS-associated symptoms (Zhou, Zhang, and Verne 2009). These findings have also been confirmed in animal experiments. In rats, it has been shown that increased intestinal permeability triggered by stress increase visceral sensitivity. In rats, intestinal permeability alterations have also been attributed to hormonal fluctuations throughout the menstrual cycle (Ait-Belgnaoui et al. 2005; Braniste et al. 2009).

There are gender differences concerning symptoms of IBS, also women of productive age are more prone to develop IBS than men, and symptoms are more severe in them (Österberg et al. 2000; Thompson et al. 2000). As animal studies have shown, abdominal pain, a hallmark of IBS, has variations across the menstrual cycle, aggravated at menses compared to other cycle periods. As a premenstrual drop in the level of ovarian hormones coincides with the intensification of IBS symptoms, ovarian steroids seem to be a protective factor against exacerbation of premenstrual IBS pain (Smith et al. 2006; Houghton et al. 2002).

Administration of estradiol to ovariectomized rats can reduce intestinal permeability by enhancing epithelial tight junctions. Therefore, it can be concluded that the high level of estrogen during the follicular cycle of the menstrual cycle reduces intestinal permeability and mitigates the symptoms of IBS (Braniste et al. 2009). Another finding confirming these statements is the expression of estrogen receptors (ERs) in the gastrointestinal tract with the dominance of ER b receptor in the colon mucosa (Wada-Hiraike et al. 2006). Regarding animal and human studies, estrogen is involved in epithelial barrier development of the colon by activating ER b so that it can decline colon permeability (Wada-Hiraike et al. 2006; Braniste et al. 2009).

Soybean contains daidzein, glycitein and genistein isoflavones present in aglycone forms by fermentation which can be absorbed through the intestine

(Hubert et al. 2008; Setchell et al. 2002). Due to the structural similarity to 17b-estradiol, soy isoflavones are considered phytoestrogens with a high affinity to ER b (Morito et al. 2001; Birk 1985). The inhibitory effects of soy on stress-related symptoms are similar to those of estradiol, which is attributed to the preventive effects of soy isoflavones on ER. Therefore, the daily intake of soy by women of childbearing age can prevent stress-induced hyperpermeability of the intestine and visceral hypersensitivity (Moussa et al. 2013).

2.2.4. Inflammatory Bowel Disease (IBD)

Inflammatory bowel disease (IBD) is defined as chronic inflammation of the colon and/or small intestine, although it may involve any part of the gastrointestinal tract. This disorder is divided into ulcerative colitis (UC) and Crohn's disease (CD). No definitive cure is known for IBD at present. This has led to the requirement for alternative treatment with high efficacy and minimal toxic effects. No epidemiological study has been published on the association of polyphenols with inflammatory bowel disease, although several studies indicate the reverse association between IBD and fruit and vegetable consumption (Amre et al. 2007; Hou, Abraham, and El-Serag 2011). Furthermore, there is a lack of direct evidence from clinical trials to address polyphenols' efficacy in patients with IBD. Hence our limited knowledge about IBD is stem from limited pilot studies on the effects of curcumin effectiveness for UC or CD (Hanai et al. 2006; Suskind et al. 2013; Holt, Katz, and Kirshoff 2005), bilberries for UC (Biedermann et al. 2013), or ECGC for UC (Dryden et al. 2013), which almost all of them have had promising outcomes. Experimental studies in this regard are broader than human studies. Histological findings in a study conducted on mice suffering from colitis revealed an ulceration reduction in the mucosa and submucosa of the colon due to a 5% black raspberry diet (Wang et al. 2013). Another study on mice with ulcerative colitis disclosed that consumption of cocoa polyphenols could be effective in preventing weight loss and reducing the need for resection of affected colon segments (Andújar et al. 2011).

In an estimation of dietary intake of proanthocyanidins, the polymers of flavan-3-ols, in US adults, it was shown that the grape seed polyphenol in all quantities could avert weight loss and visible damage. Proanthocyanidins with the preservation of superoxide dismutase and glutathione peroxidase activity are capable of strengthening the antioxidant capacity of the colon. Reducing TNF-α and NF-κB levels is another effect of this polyphenol (Wang, Ge, et al. 2011).

Various studies have highlighted the effectiveness of pure polyphenol or natural plant extracts and introduced them as a novel treatment (Kaulmann and Bohn 2016; Farzaei, Rahimi, and Abdollahi 2015). It seems that one of the most effective compounds in this field is curcumin, which can be used as a safe and promising choice for the remission of IBD (Hanai et al. 2006). In addition to curcumin, other polyphenols, including resveratrol, quercetin and anthocyanin, have also had promising results in improving colitis in animal models. Unfortunately, these results have not been widely studied in human interventional studies, so evidence remains limited in this regard. However, the results of pre-clinical studies suggest that polyphenols can be used as anti-colitis therapy (Martin and Bolling 2015).

In interpreting the results of animal experiments and generalizing it to humans, it should be noted that the metabolism of bioactive compounds and gut microbiota composition in animal models are dissimilar to humans (Mackenzie et al. 2005; Nguyen et al. 2015). It is important to note that IBD is associated with alterations in microbiota and metabolism (Hansen et al. 2012; Nell, Suerbaum, and Josenhans 2010; Zhang et al. 2012), and consumption of bioactive compounds, such as polyphenols, can modify the microbiota composition (Cowan et al. 2014; Espley et al. 2013; Etxeberria et al. 2013).

The promising effects of polyphenols reported in animal studies are reducing bleeding, improving fecal consistency, improving histological findings, preventing weight loss, and reducing the need for bowel resection. Since polyphenols can suppress inflammatory responses, the same results can be achieved in humans by applying the appropriate dose of polyphenols. Reducing inflammatory cytokines, inhibiting NO production and COX-2 activity is among these effects (Martin and Bolling 2015).

EGCG is one of the polyphenols studied frequently in IBD animal models. Reducing the severity of diarrhea, preventing weight loss and improving inflammatory parameters include the beneficial effects of EGCG (Abboud et al. 2008). There is some evidence; however, that high doses of EGCG may exacerbate inflammation and deteriorate colitis while it can be beneficial at low amounts without side effects (Guan et al. 2012). Therefore, patients with IBD should be avoided from taking supplements containing high doses of this bioactive substance until further studies on the efficacy and safety of EGCG have been performed (Martin and Bolling 2015).

For several reasons, polyphenols can be used as an adjuvant to current IBD drugs. First, the effect of polyphenols on TNF-α reduction has been proven in various in vivo and in vitro studies and one of the IBD treatment approaches is the use of drugs that inhibit TNF-α. In addition to mitigating the inflammatory response as a mechanism for the anti-colitic, polyphenols can effectively improve the function of the intestinal barrier, whose role has been proven in IBD (Turner 2009; Azuma et al. 2013; Suzuki 2013; Suzuki and Hara 2011). These broad actions of polyphenols are worthy of further investigation. Figure 4.1 presents the underlying mechanisms of action of anthocyanins in the management of IBD.

The main cause of attention given to soy in recent years is probably its anti-inflammatory properties, which has been attributed to the reduction of the expression of inflammatory mediators, based on animal studies (Seibel et al. 2009; Young et al. 2011). The properties described for soy and soy extracts are mainly due to the presence of isoflavones, including genistein, daidzein and its metabolite, named equol, which exhibits estrogen-like activity (Kuiper et al. 1998).

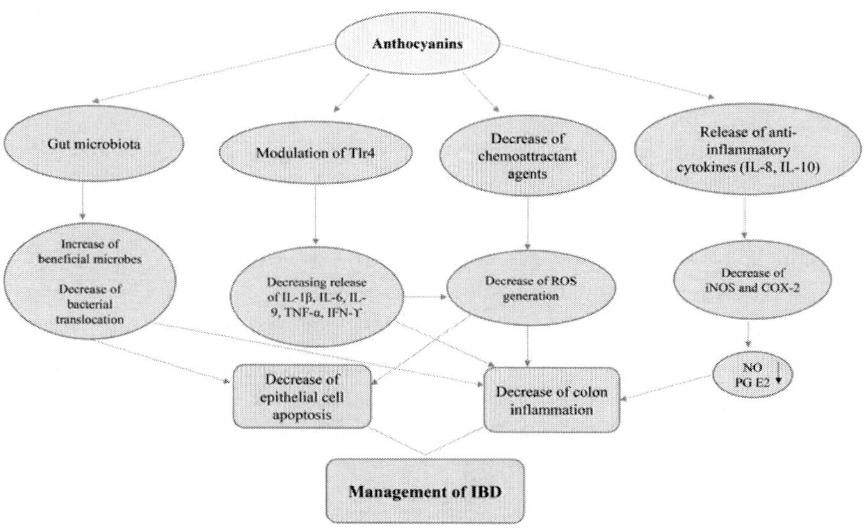

Figure 4.1. Mechanisms of the effect of anthocyanins in management of IBD. IBD: Inflammatory bowel disease; iNOS: Inducible nitric oxide synthase; COX-2: Cyclooxygenase 2; NO: Nitric oxide; PG: Prostaglandin; ROS: Reactive oxygen species; Tlr: Toll-like receptor.

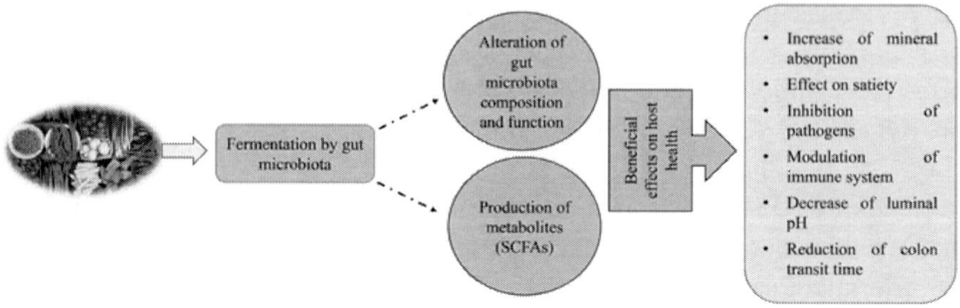

Figure 4.2. Beneficial effects of prebiotics on health.

Although the exact mechanism of soybean effects on IBD is still unclear, it seems that soy isoflavones exert anti-colitis effects through binding to estrogen receptors (ERs) in the colon. In this way, the binding of isoflavones to the ER enhances the expression of transmembrane tight junction proteins (Braniste et al. 2009) and suppresses the production of inflammatory cytokines (Houdeau et al. 2007; Verdú et al. 2002). Thus, the isoflavones can improve the integrity and function of the intestinal barrier (Suzuki and Hara 2011; Seibel et al. 2009; Morimoto et al. 2009). Given the association of gut barrier dysfunction and IBD, considering these mechanisms of isoflavones actions provide a rationale for the potential use of this compound as adjuvant therapy in IBD (Moussa et al. 2012).

Since there is an accumulating body of evidence confirming the correlation between intestinal microflora imbalance and IBD, on the other hand, the complication of this imbalance, such as inflammation is well documented, the use of probiotics, prebiotics and their combination termed synbiotic can have a great effect on IBD treatment (Camilleri 2006). The association between gut microflora is revealed with many diseases, including IBD, and its imbalance can be a stimulant for the onset of the diseases (Chandran et al. 2003). Hence, one of the most common uses of probiotics is the treatment of inflammatory diseases, including Crohn's disease (Schultz et al. 2004; Prantera et al. 2002) and ulcerative colitis (Rembacken et al. 1999; Bibiloni et al. 2005), although the precise mechanism of probiotic action is still not well known. *Mycobacterium paratuberculosis* is a microorganism that is involved in the onset and progression of IBD. In more than 80% of patients with colitis, an antibody has been identified against *Mycobacterium avium paratuberculosis* spp., which is also seen in intestinal biopsies of patients with IBD (Greenstein 2003). These bacteria have been shown to exert adverse effects on the metabolism of fatty acids and SCFAs by producing hydrogen sulfide, disrupting the secretion of the protective mucus in the intestinal epithelium and intensifying inflammation (Bäckhed et al. 2005).

The imbalance of the intestinal microflora population in the form of an increase in the number of aerobic bacteria such as *Escherichia coli* and anaerobic bacteria, such as *Bacteroides*, and the reduction of favorable microorganisms, such as *Lactobacillus* and *Bifidobacterium*, are reasonable rationale to use probiotics in IBD therapy (Swidsinski et al. 2002). In a study conducted in 2013 in patients with IBD, Shadnoush and colleagues showed that consumption of probiotic yogurt containing *Lactobacillus* and *Bifidobacterium* significantly reduced the serum IL-1β, TNF-α, and C-reactive protein level after eight weeks compared to plain yogurt, while IL- 6 and IL-10 significantly increased (Shadnoush et al. 2013). These results indicate that probiotics can contribute to the maintenance of gastrointestinal hemostasis and normalize the inflammatory responses of intestinal immunocytes. In another study, daily consumption of 100 mL of fermented milk containing *Bifidobacterium* for one year could prevent UC relapses and help sustain its remission (Ishikawa et al. 2003).

One of the most common probiotics preparations is VSL#3, which contains eight species, namely *Lactobacillus casei*, *Lactobacillus plantarum*, *L. acidophilus*, *Lactobacillus delbrueckii* subsp. *bulgaricus*, *Bifidobacterium longum*, *Bifidobacterium brevis*, *Bifidobacterium infantis* and *Streptococcus thermophilus*. This combination has a proven effect in reducing inflammation (Lammers et al. 2003). Studies have also shown that VSL#3 can effectively enhance the integrity of the intestinal epithelial barrier, as it can increase the expression of tight junction proteins and reduce apoptosis of the epithelial cells (Mennigen et al. 2009). Administration of VSL#3 in children caused and maintained remission by more than 90%, compared with the placebo group (Miele et al. 2009). Similar results have been reported in adults (Shen, Zuo, and Mao 2013). Prescription of this probiotic preparation increases the gastrointestinal bacterial diversity in IBD patients (Kuehbacher et al. 2006).

Another combination that can be suitable in balancing the microflora of the intestine is prebiotic, which has been exposed to be effective in treating and preventing the recurrence of enteritis. Prebiotics are indigestible components of the diet that undergo bacterial fermentation in the large intestine (Figure 4.2). The metabolizing of oligosaccharides to SCFA can selectively enhance beneficial bacteria's growth and activity, prevent pathogens, and ultimately improve host status (Gibson and Roberfroid 1995). Reducing intraluminal pH under the influence of prebiotics can also have bifidogenic effects (Cummings and Macfarlane 2002; Roberfroid and

Delzenne 1998) and prevent diarrhea by inhibiting the growth of strains of pathogenic bacteria, e.g., *Clostridium* spp. (Cummings and Macfarlane 2002). Despite numerous reports on the affirmative effects of FOSs on increasing bifidobacteria and decreasing *Bacteroides* and *Clostridium* populations (Gibson et al. 1995), some studies fail to show such an effect in patients with Crohn's disease (Benjamin et al. 2011).

Improving the structure and function of the intestinal mucous membrane is another mechanism of prebiotic action (Muccioli et al. 2010). This property of prebiotics has been attributed to their role in the production of SCFA and the stimulation of the synthesis of proteins forming tight junctions. Furthermore, the role of prebiotics in reducing inflammatory mediators is also essential (Langen, Mirjam, and Dieleman 2009; Smith et al. 2013).

Given the benefits of probiotics and prebiotics in the treatment of IBD, it seems that simultaneous consumption of both compounds, named synbiotics, can exert synergistic effects. Synbiotic improve the growth of favorable bacteria and prevent the growth of pathogens. To confirm this, Bomba and colleagues in 2002 showed that administration of *Lactobacillus paracasei* and FOS could increase the *Lactobacillus* and *Bifidobacterium* population and reduce *Clostridium* and Enterobacterium (Bomba et al. 2002). Reducing inflammation in biopsy samples from US patients combined with *B. longum* has also been reported (Furrie et al. 2005; Steed et al. 2010). In addition, different positive effects, including improving micronutrient absorption such as calcium, magnesium and phosphorus, have also been reported for synbiotic, which is of particular importance in patients with IBD (Rodrigues et al. 2012).

2.2.5. Celiac Disease (CD)

Experimental studies on intestinal and circulating cells propose an association between oxidative damage and celiac disease. Disruption of antioxidant defense by inducing inflammation to the mucosa causes the cells to be vulnerable to oxidative damage. On the other hand, interrupts the recovery and repairing of the integrity of the epithelial cells (Ferretti et al. 2012).

Reducing the production of inflammatory mediators, decreasing oxidative stress, promoting anti-inflammatory responses through effects on cell signaling and expression of genes, and improving the function of the gut barrier are among the targeted mechanisms for the treatment of celiac disease (Ferretti et al. 2012). Various dietary components, including plant polyphenols and fatty acids, can be effective in the nutritional treatment of celiac disease through modifying the mentioned mechanisms (Calder et al. 2009; Calder 2011).

Various dietary components, including polyphenols such as resveratrol, quercetin, hesperidin and curcumin, possess anti-inflammatory and anti-oxidative effects that can have protective effects on the intestinal epithelium (Calder et al. 2009). Accordingly, including these components in the diet of patients with celiac disease can be a therapeutic opportunity by maintaining intestinal barrier integrity and playing a protective role against toxicity of gliadin peptide (Matysiak-Budnik et al. 2003).

Fruits and vegetables with a variety of polyphenols and carotenoids can produce beneficial biological effects on the epithelial cells of the intestine. Indeed, the major effects of these compounds are attributed to their antioxidant and anti-inflammatory properties. The potential target of these bioactive compounds appears to be the transduction signaling cascade, which leads to the triggering of transcription factors such as NF-κB (Hecker et al. 1996; Heber and Lu 2002; Di Carlo et al. 1999).

Prominent among polyphenols are flavonoids, which by suppressing the enzymes in the pathway of eicosanoid synthesis including phospholipase A2, COX and lipoxygenase (LOX), on one hand, and inhibiting the induction and expression of iNOS, on the other hand, ultimately lead to a lower concentration of prostanoids and leukotrienes and exert their anti-inflammatory activity in different cells (Di Carlo et al. 1999; Calder et al. 2009). Inhibiting the expression of the iNOS and COX-2 gene has been occurred by preventing the activation of NF-κB, interferon regulatory factor 1 (IRF-1) and Signal transducer and activator of transcription *1* (STAT-1α) at the transcriptional level. This is associated with inhibiting the production of ROS generation due to gliadin and IFN-γ. Studies have shown that the expression of iNOS and COX-2 gene is induced by gliadin via stimulation of IFN-γ, and bioactive compounds such as lycopene and quercetin can reduce this gene expression (De Stefano et al. 2007).

In addition to preserving the function of the gut barrier, intestinal tight junction also plays a crucial role in the paracellular transmission of gluten. Oxidative stress and inflammatory cytokines cause intestinal gut barrier dysfunction that flavonoids can ameliorate. As a member of the flavonoids family, quercetin can help treat celiac disease by enhancing the function of intestinal tight junction through the assembly (by changing in phosphorylation status) and expression of the intestinal tight junction proteins. The protective effects of other flavonoids, including epigallocatechin gallate, genistein, myricetin, have also been reported on intestinal tight junction and gut barrier function (Suzuki and Hara 2011).

Flavonoids' anti-inflammatory and antioxidant role has been reported abundantly in various studies. Inhibiting the expression of inflammatory proteins and enzymes using carotenoids is partly due to the suppression of NF-κB activation. This effect is mediated by inhibiting various protein kinases involved in the cell signaling pathway, such as mitogen-activated protein kinase and extracellular signal-regulated kinase (Karell et al. 2003; Heber and Lu 2002). This mechanism can justify the efficacy of flavonoids in celiac disease management.

In celiac disease, akin to other gastrointestinal disorders, imbalance in intestinal microbiota has been abundantly reported (Nadal et al. 2007; De Palma et al. 2010; Collado et al. 2008). The abnormal composition of intestinal microbiota is characterized by the reduction of favorable bacteria, such as *Bifidobacterium* spp., and the increased population of pathogens, including *Bacteroides* spp., which is found in duodenal biopsy of untreated patients (Nadal et al. 2007; Collado et al. 2008). Furthermore, the association of intestinal microbiota composition with clinical manifestations of celiac disease has been proven and can vary according to the severity of the symptoms (Pirjo Wacklin et al. 2013). It has also been observed that after following the gluten-free diet and normalization of the morphology of small intestine mucosal cells, gastrointestinal microbiota dysbiosis persists in patients with celiac disease (Collado et al. 2008; Nadal et al. 2007; Nistal et al. 2012). Therefore, lack of adequate response to treatment as well as the persistence of gastrointestinal symptoms can be attributed to the imbalance in intestinal microbiota in treated celiac patients (Wacklin et al. 2014). Intestinal microbiota disturbance is independent of the fecal pH. Supplementation with probiotics may increase intestinal bifidobacteria, which also appear in the feces (Martinello, Roman, and Souza 2017).

One of the most common and complicated complications in celiac patients is a marked reduction in bone mineral density (BMD). Usually, interventions with calcium and vitamin D supplementation have little success in increasing BMD. Failure to respond to calcium and vitamin D supplementation is attributed to the absence of sufficient amounts of calbindin (calcium- binding protein) in the small intestine of patients with celiac disease (Capriles, Martini, and Arêas 2009). One of the novel therapeutic approaches to increasing BMD and improving the absorption of calcium from the intestine is prebiotics, especially inulin and oligofructose. Inulin type fructans are soluble dietary fiber, which is considered a prebiotic. Prebiotics are fermentable food components that, by improving the composition and activity of the intestinal microbiota, confer unique health benefits to the host (Abrams et al. 2005;

Abrams et al. 2007; Gibson et al. 2004). Around 70% of the increased calcium absorption due to fructans consumption is attributed to colonic absorption (Abrams et al. 2007). This effect becomes important in gastrointestinal absorption disorders such as celiac disease (Abrams et al. 2007; Capriles, Martini, and Arêas 2009).

2.3. Liver

Hepatic disorders mainly include non-alcoholic fatty liver disease and cirrhosis discussed in detail in Chapter 1.

2.3.1. Non-Alcoholic Fatty Liver Disease (NAFLD)

The increasing prevalence of non-alcoholic fatty liver disease and the lack of definitive treatment for the disease are among the reasons that have led patients to take non-pharmacological therapies.

Today, the pathogenesis of the non-alcoholic fatty liver disease is justified based on the multifocal hypothesis mentioned in the previous chapter. Polyphenols can be considered as a potential treatment for NAFLD and prevent progression to non-alcoholic steatohepatitis (NASH) (Akhlaghi 2016; Rodriguez-Ramiro, Vauzour, and Minihane 2016) because they target two second-hits, inflammation and oxidative stress. Supplementation with polyphenols can maintain the homeostasis of the redox system and prevent the production of active oxygen species (ROS) (Ying et al. 2013).

One of the important effects of polyphenols is suppressing inflammation (Rodriguez-Ramiro, Vauzour, and Minihane 2016). Inflammation by activating collagen production in stellate cells causes hepatic damage leading to fibrosis (Marra et al. 2008). Various animal and experimental studies have been conducted to confirm the beneficial effects of different polyphenols on oxidative stress (Gómez-Zorita et al. 2012; Kobori et al. 2011), inflammation (Kobori et al. 2011; Guo et al. 2012) and consequently hepatic steatosis (Zhang et al. 2011; Wang et al. 2009; Alberdi et al. 2013).

One of the polyphenols whose anti-inflammatory and anti-oxidative properties may offer a potential therapy for NAFLD and be beneficial in preventing the progression of steatohepatitis is resveratrol (Heeboll et al. 2016). Most effects of resveratrol are exerted through activation of AMP-activated protein kinase (AMPK) and silent information regulation 2 homolog 1 (SIRT1), which play a central role in the metabolism of various cells (Zang et al. 2006; Smith et al. 2009). The investigations indicate that resveratrol, by

AMPK/SIRT1-mediated inhibition of lipogenic enzymes, can ultimately attenuate hepatic lipogenesis (Alberdi et al. 2013; Li et al. 2011; Baur et al. 2006). In addition to reducing the synthesis of fatty acids, resveratrol also increases the β-oxidation of fatty acids by stimulating mitochondrial biogenesis through PGC-1α activation and up-regulates hepatic expression of uncoupling protein 2 (Morten Møller Poulsen et al. 2012). Further research is required to prove this hypothesis that taking resveratrol with a high-fat diet can prevent hepatic fat accumulation through increasing the number of mitochondria and uncoupling protein and, consequently, dissipating excess energy as heat (Morten Møller Poulsen et al. 2012). Furthermore, resveratrol has also been referred to modulate the expression of fatty acid transporters (Ahn et al. 2008; Xin et al. 2013). Based on studies conducted in animal and laboratory models, an effective dose of resveratrol has been reported in the range of 10-50 μmol/L to produce anti-lipogenic effects that can reduce de novo lipogenesis (Aguirre et al. 2014).

Major mechanisms of resveratrol action in reducing liver inflammation include reducing the infiltration of macrophages into the hepatocyte (Jeon et al. 2012), decreasing serum TNF-α levels (Bujanda et al. 2008; Li et al. 2014), and inhibiting the NF-κB pathway (Li et al. 2014; Do et al. 2012). Another highly desirable effect of resveratrol is the improvement of insulin sensitivity in peripheral tissues (Lagouge et al. 2006; Szkudelski and Szkudelska 2011), which has been related to its potentially positive systemic effects. Except for direct effects on liver fat accumulation, resveratrol also can improve the metabolic risk factors and allied clinical symptoms, including dyslipidemia, insulin resistance, increased levels of inflammatory cytokines and liver enzymes (Faghihzadeh, Adibi, and Hekmatdoost 2015; Heebøll et al. 2016; Chen et al. 2015; Zhang, Yuan, et al. 2016). However, the clinical trials of resveratrol on NAFLD have contradictory results (Morten M Poulsen et al. 2013; Faghihzadeh et al. 2014; Chachay et al. 2014; Chen et al. 2015). The inconsistency of the results between studies may stem from the number of factors, including different patients' baseline characteristics, intervention period, a dosage of resveratrol, the severity of the disease, and so on. Taken together, what can be deduced from various studies is that the effect of resveratrol in the prevention of NAFLD is superior to its effect on treatment. Second, resveratrol is more effective in a NAFLD model of mild steatosis, not established NASH. Finally, the lack of response to resveratrol can be attributed to the low bioavailability of this polyphenol (Patel et al. 2013; Heeboll et al. 2016).

Hesperidin is a kind of polyphenol from the flavanone family that has recently shown promising results in fatty liver. In addition to antioxidant activity, hesperidin can favorably modulate lipid metabolism by altering the expression of related proteins. This effect is evidenced by increased β-oxidation of fatty acids and lipid synthesis inhibition (Wang, Hasegawa, et al. 2011; Constantin et al. 2013; Nichols et al. 2011). Therefore, hesperidin can be considered a therapeutic option for hepatic steatosis and fibrosis by reducing liver fat content and diminishing hepatic inflammation and oxidative stress (Jeon et al. 2014; Elshazly and Mahmoud 2014; Pérez-Vargas et al. 2014).

Quercetin, a member of polyphenolic compounds, has been known for its potent antioxidant properties (Susana Martínez-Flórez et al. 2005; Crespo et al. 2008). Quercetin has been stated to exhibit a wide range of biological effects that are found to be attributed to its anti-inflammatory and antioxidant activity, including beneficial effects on cancers (Bischoff 2008; Chirumbolo 2010; Jung et al. 2013) and cardiovascular diseases (Jung et al. 2013; Perez-Vizcaino et al. 2009; Han et al. 2009). Experimental evaluations also confirm the therapeutic effects of quercetin in hepatic injury due to its antioxidant activity in hepatocytes (Cuevas et al. 2011). Another mechanism of quercetin action in fatty liver is the reduction of liver lipid accumulation through inhibition of fatty acid omega-oxidation (Hoek-van den Hil et al. 2013). The effect of quercetin on attenuating hepatic steatosis was first described in 2011 in animal models, that quercetin could improve insulin sensitivity and increase plasma adiponectin concentrations (Kobori et al. 2011). One year later, Paul and colleagues reported that quercetin prevents liver damage by inhibiting NF-κB activation in its downstream pathways in rats fed with high-fat and high-carbohydrate diets (Panchal, Poudyal, and Brown 2012). Another finding from animal model investigations is the inhibition of hepatic stellate cell activation that correlates with the reduction of liver fibrosis (Lang et al. 2001; Hernández-Ortega et al. 2012). The results of laboratory studies also indicated that treatment of HepG2 cells with 10 μmol/L quercetin for 24 h reduced the accumulation of triglycerides, insulin resistance, and inflammatory cytokine production. Accordingly, the researchers concluded that quercetin could effectively improve the symptoms of NAFLD (Vidyashankar, Varma, and Patki 2013). Part of these effects of quercetin is expected to be mediated by the inactivation of PI3K/AKT pathway. Quercetin treatment in both *in vivo* and *in vitro* models of NAFLD can regularize the expression of genes associated with inflammation, oxidative stress and lipid metabolism. Hence, quercetin can be a potential strategy to prevent the development and progression of NAFLD (Pisonero-Vaquero et al. 2015). However, human studies in this area remain very limited.

The major part of soy's beneficial health effects is attributed to its isoflavones content. However, it should be noted that the major evidence in this regard is derived from studies conducted on animal models. Feeding obese Zucker rats with a diet containing soy protein with isoflavones improved serum lipid profiles and reduced fat accumulation in the liver (Peluso et al. 2000). Similar results have also been obtained in another study (Wergedahl et al. 2004). It has been shown that the accumulated fat droplets in the liver of isoflavone-supplemented rats were fewer and smaller than that of the control group. The quantitative amount of liver triglyceride also was decreased subsequently. The proposed mechanisms for health effects of isoflavones include increased β-oxidation of fatty acids, reduced lipid synthesis, enhanced hepatic fatty acid catabolism and decreased VLDL clearance from serum by the liver (Gudbrandsen et al. 2006).

Stimulating fatty acids oxidation by genistein, a soy isoflavone, is mediated by increasing PPAR-α activity and downstream signals (Kim et al. 2004; Qiu and Chen 2015). On the other hand, genistein suppresses de novo lipogenesis in the liver by inhibiting sterol regulatory element-binding protein-1 (SREBP-1) and its targets (Shin et al. 2007). These findings were confirmed in an animal study in which subcutaneous injection of genistein led to decreased steatosis and inflammation in the liver (Yalniz et al. 2007). Dietary supplementation of genistein also had similar effects (Kim, Kang, and Lee 2010). Daidzein, other isoflavones of soy, may exert insulin- sensitizing and anti-steatosis effects through downregulating the liver X receptor (LXR) and carbohydrate response element binding protein (ChREBP); (M. Kim et al. 2011). Furthermore, by up-regulating the expression of UCP-1, daidzein promotes thermogenesis in brown adipose tissue that is followed by improved NAFLD (Crespillo et al. 2011). Moreover, reducing inflammation and oxidative stress are considered two of the most favorable effects of soy isoflavones (Qiu and Chen 2015).

Notch is one of the important signaling pathways related to hepatic fat accumulation, which is considered a crucial regulator of homeostasis in energy (Gridley and Kajimura 2014). Curcumin is one of the most effective polyphenols in suppressing this pathway, improving insulin sensitivity and relieving high fat diet-induced hepatic steatosis. Improving lipid profile, reducing central obesity, and improving glucose hemostasis are other favorable effects reported for curcumin (Zhao et al. 2018). In addition to the NAFLD, Notch signaling has been seen in obesity and diabetes. The activation of this pathway is responsible for promoting hepatic lipogenesis and fat production, which ultimately causes hepatosteatosis (Kim et al. 2014). Curcumin also improves contributing risk factors associated with NAFLD, including high plasma triglyceride and cholesterol levels (Mohammadi et al. 2013; Sahebkar 2014b, 2014c; Sahebkar, Chew, and Watts 2014), insulin resistance (Sahebkar 2013; Na et al. 2013; Chuengsamarn et al. 2012), mitochondrial dysfunction (Ramirez-Tortosa et al. 2009), increased adipogenesis (Bradford 2013), pro-inflammatory status (Panahi et al. 2012; Sahebkar 2014a) and oxidative stress (Sahebkar et al. 2013; Panahi, Alishiri, et al. 2016; Panahi, Ghanei, et al. 2016). One of the promising outcomes of curcumin supplementation is its potential role in reducing hepatic lipid accumulation and preventing diet-induced NAFLD from advancing to NASH (Hasan et al. 2014; Öner-İyidoğan et al. 2013).

Green tea extract has a long history of treating liver disorders due to catechins (Imai and Nakachi 1995). Green tea extract may mitigate the multiple hits of NAFLD through hypolipidemic effects (Babu, Pon, and Liu 2008) and antioxidant activity (Frei and Higdon 2003). The expression of hepatic lipogenic genes including SREBP-1c and fatty acid synthase (FAS) in fructose-fed ovariectomized rats treated with green tea extract decreased (Shrestha et al. 2009). Consumption of green tea extract in obese mice prevented steatosis and hepatic injury (Bruno et al. 2008). Green tea, due to its content of catechins, exhibits protective effects by reducing the peroxidation of lipids and ROS production (Nakamoto et al. 2009). The results of the animal study in this regard suggest that Epigallocatechin gallate (EGCG), the principal green tea catechin, can reduce the accumulation of fat in the liver and adipose tissue by reducing the expression of adipogenic genes (Bose et al. 2008; Lee, Kim, and Kim 2009). Another effect of EGCG is to prevent hepatotoxicity by reducing inflammation and lipid peroxidation (Chen et al. 2004). Green tea consumption has hepatoprotective activities through reducing the expression of hepatic PGE2 and COX-2 and can prevent NAFLD from progressing to NASH (Chung et al. 2015; Imai and Nakachi 1995). However, in some studies, it has been reported that consumption of polyphenol-rich green tea extract in rats and a high-fat diet and/or hypercholesterolemia can lead to steatosis and hepatic injury was attributed to polyphenol-induced inflammation and oxidative stress. In addition, green tea extract can alter the pathway of bile acid synthesis and cholesterol clearance (Hirsch et al. 2016). These findings contradict previous studies in which green tea extract was found to reduce oxidative stress and inflammatory responses accompanying NAFLD development due to its polyphenols content, and it was even mentioned that these effects of the green tea extract are dose-dependent. Green tea extract induces protective effects against lipid peroxidation and liver damage using restoring glutathione levels and stimulating antioxidant enzymes activity. Reducing the expression of TNF-α in the liver and adipose tissue, which have increased in NAFLD, is another protective mechanism of green tea extract. A study also revealed that green tea extract reduced the transfer of free fatty acids to the liver by decreasing lipogenesis and lipolysis in adipose tissue (Park et al. 2011). Collectively, hepatic lipid peroxidation inhibition, antioxidant defenses enhancement, and pro-inflammatory responses reduction are mechanisms by which polyphenol-rich green tea may mitigate hepatic injury in a dose-dependent manner (Park et al. 2011).

Carotenoids are bioactive compounds found in fruits and vegetables abundantly. Numerous studies have been conducted on the effects of carotenoids, especially lycopene and β-carotene, on human health. Part of these researches are related to the antioxidant and anti-inflammatory effects of carotenoids on NAFLD; however, the precise mechanism of protective effects of carotenoids is unknown (Rao and Rao 2007; Kaulmann and Bohn 2014). Association studies of NAFLD and carotenoids have shown that the higher the fat accumulation in the liver, the lower the serum levels of β-carotene, but the levels of lycopene, α carotene, β-cryptoxanthin, and lutein remained constant (Park et al. 2008). Similar results have been reported in patients with NASH that most of the carotenoids (lutein, zeaxanthin, β-cryptoxanthin, lycopene, α-carotene, and β-carotene) is lower than that of healthy subjects (Erhardt et al. 2011). On the other hand, as the disease progresses, serum levels of carotenoids appear to decrease and vice versa (Cao et al. 2015). In an animal study conducted in 2010 on rats fed a high-fat diet, the preventive role of lycopene in NASH has been mentioned (Bahcecioglu et al. 2010). The mechanism of lycopene in hepatic steatosis prevention seems to be through reducing inflammation, which was manifested by lower serum levels of MDA and TNF-α and higher hepatic GSH levels. In fact, in this study, high-fat diet-induced NASH, oxidative stress and inflammation in rats, and lycopene administration had positive effects on prevention and treatment (Bahcecioglu et al. 2010). Similar results have been acquired with the consumption of tomato juice in animal models (Martín-Pozuelo et al. 2015; Bernal et al. 2013). It should be noted that absorption of carotenoids is dependent on lipid and the absorption of these compounds improves with dietary fat (Unlu et al. 2005). The role of beta carotene has been shown to relieve hepatic inflammation, fibrosis and cirrhosis, which may have both preventive and therapeutic effects

(Seifert et al. 1995; Wardi et al. 2001). In addition to supplementation of different carotenoids in the form of purified and extract, studies have shown that the dietary sources of carotenoids include apricots, tomatocs, and so on, also have similar effects (Ozturk et al. 2009; Tainaka et al. 2011). In addition to NAFLD, these bioactive compounds can also ameliorate obesity, dyslipidemia and insulin resistance. Increasing the activity and expression of antioxidant enzymes including catalase, GSH peroxidase and superoxide dismutase, decreasing the expression of lipogenic enzymes and stimulating transcription factors involved in fatty acid oxidation are among the mechanisms proposed for these effects (Ozturk et al. 2009; Chaves et al. 2008).

Flaxseed is one of the richest sources of phytoestrogen, a member of the polyphenol family. Furthermore, flaxseed is a good source of fat (41%), protein (20%), dietary fiber (28%) and mineral-rich residue (Weill et al. 2002). More than half (57%) of flaxseed fat content is α-linolenic acid (ALA), which makes it the richest source of ALA, with 16% linoleic acid (Commission 2001). The fiber content of the seed is a mixture of soluble and insoluble fiber. Cellulose, the main ingredient of the cell wall structure; lignin, the cell wall of woody plants and mucilage gum is the most important flax dietary fiber products (Friedman and Levin 1989). The other bioactive compound in the seed, lignan, plays a significant role in human health as a phytochemical agent, particularly cancer prevention (Aubrecht et al. 1998). Lignans are phenolic compounds with apparent antioxidant potential. Lignans of the flaxseed called secoisolariciresinol diglycoside (SDG), which are subjected to bacterial metabolism in the colon (Jones, Lineback, and Levine 2006; Morris 2007).

Although numerous health-beneficial effects have been described for flaxseed, only one clinical trial examined its therapeutic effects on NAFLD to the best of our knowledge. The findings of a study indicated that flaxseed supplementation in combination with a lifestyle modification could be considered adjuvant therapy for patients with NAFLD through ameliorating insulin resistance, dyslipidemia and inflammation (Yari et al. 2016). However, in previous animal studies, the benefits of flaxseed consumption were shown in improving oxidative stress and preventing the hepatic accumulation of fat (Kelley et al. 2008; Xu et al. 2013). In addition, omega-3 fatty acids have tremendous benefits, and flaxseed is also a rich source of these fatty acids. Improving insulin sensitivity (Xu et al. 1999) and stimulating lipid β-oxidation through regulation of hepatic peroxisome proliferator-activated receptor (PPAR) alpha or gamma and sterol regulatory element- binding protein-1c (SREBP-1c) genes are the mechanisms responsible for flaxseed benefits in improving NAFLD (Marx et al. 2004; Jump 2008). In a 12-week trial, it was shown that daily supplementation of 30 g of flaxseed could significantly increase the level of total erythrocyte n-3 PUFAs, α-linolenic acid, eicosapentaenoic acid, and docosapentaenoic acid (Zong et al. 2013).

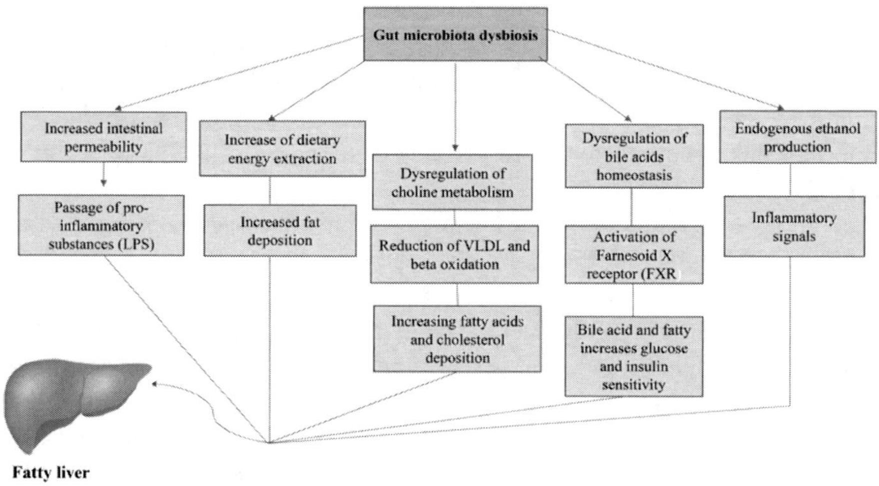

Figure 4.3. Relation between gut microbiota dysbiosis and fatty liver disease.

Gut microbiota is considered an environmental element in the pathogenesis of NAFLD (Duseja and Chawla 2014; Vos 2014) (Figure 4.3). Identifying the importance of gut microbiota composition in the development and progression of NAFLD, based on animal and human studies, has led to intensive studies of probiotics in recent years (Compare et al. 2012; Bhathena et al. 2013). Correspondingly, a direct correlation of intestinal microbiota misbalancing with increasing levels of liver enzymes, including aminotransferases in the NAFLD has also been revealed (Aller et al. 2011; Shavakhi et al. 2013). The modification of the intestinal microbiota in the form of increased the ratio of *Bifidobacterium* to *Enterobacteriaceae* in high-fat diet-fed mice was associated with reduced liver steatosis, weight loss and improved insulin sensitivity (Cano et al. 2013). In addition to the anatomical connection between the intestine and liver through the hepatic portal system. Although the mechanism of association between gastrointestinal microbial and NAFLD has not yet been fully elucidated, several mechanisms have been proposed to justify this relationship and interaction. Here are some points to keep in mind in this regard (Compare et al. 2012; Nicolucci 2015):

1. LPS is one of the most important causes, to be mentioned later.
2. The imbalance of the bacterial population in the intestine increases endogenous ethanol production, which has several side effects such as inflammatory cytokine secretion and inducing choline deficiency.
3. Choline deficiency in NAFLD may also be due to increased metabolism for the synthesis of lipoproteins (VLDL mainly) and liver lipid removal.
4. Bacterial imbalance can lead to obesity by altering energy metabolism and increasing fat storages.
5. Changing intestinal permeability can be a way to increase endotoxin entry and inflammation.

Probiotics can also affect host metabolism by regulating energy extraction from nutrients and modulating the genes involved in substrate utilization in the body (Vanni and Bugianesi 2009). Endotoxin levels are elevated in NAFLD patients (Farhadi et al., 2008, Alisi et al., 2010). At the same time, probiotics have been shown to improve the function of the epithelial barrier by improving the integrity of the intestinal epithelial cells and reducing its permeability, and also, to reduce the risk of endotoxemia in patients with fatty liver (Malasanos and Stacpoole 1991; Malaguarnera et al. 2010). Two major probiotics in this area are bifidobacteria and lactobacilli, which are among the main components of the microbial population in healthy individuals and produce a wide range of favorable effects on host health (Gibson and Wang 1994; McCartney, Wenzhi, and Tannock 1996).

It is well known that the higher the Gram-negative bacteria in the intestine, the greater the risk of exposure to lipopolysaccharides (LPSs), which is a main component of the Gram-negative bacterial wall. LPS is a causal factor in insulin resistance, secretion of inflammatory mediators and NAFLD development (Manco, Putignani, and Bottazzo 2010; Cani, Amar, et al. 2007). Other gut microbiota by-products other than LPS, such as ethanol and SCFAs, reach the liver through the intestinal lumen. It is well known that LPS stimulates the production of inflammatory cytokines, including TNF-α, which is considered a causative agent in the fatty liver per se. For this reason, supplementation with probiotics or synbiotics can reduce inflammation. In a study of 52 patients with fatty liver, supplementation with synbiotic preparation for 28 weeks could significantly inhibit NF-κB and reduce TNF-α production (Eslamparast et al. 2014). Although the mechanism of this action of probiotic is still not fully defined, the results are contradictory. For instance, similar studies failed to demonstrate the reduction of TNF-α following supplements with probiotics (Aller et al. 2011; Vajro et al. 2011) and synbiotics (Alisi et al. 2014; Wong et al. 2013). These contradictions may be due to differences in the design of the trials or the bacterial strains and dosage of probiotics used.

SCFAs stimulated the synthesis and storage of triglycerides in the liver. The accumulation of triglycerides in the liver interferes with hepatic detoxification and leads to more accumulation of fat content in the liver and exacerbation of fatty liver. Therefore, it seems that consuming probiotics can be a favorable alternative therapy in patients with NAFLD by deterring fat accumulation in the liver. This hypothesis was also confirmed in a study of patients with NASH who received synbiotic (containing five probiotic strains- *L. plantarum*, *L. delbrueckii* subsp. *bulgaricus*, *L. acidophilus*, *L. rhamnosus*, *B. bifidum*- and inulin as a prebiotic) supplementation for 6 months (Wong et al. 2013).

Improvement of liver function during the consumption of *L. delbrueckii* subsp. *bulgaricus* and *S. thermophilus* was demonstrated as a reduction in liver enzymes levels including alanine amino transferase (ALT), aspartate amino transferase (AST) and gamma glutamyl transferase (GGT) (Aller et al. 2011). It also seems that consuming probiotic products has similar results. In a study conducted in 2014, it was revealed that daily consumption of probiotic yogurt with *L. acidophilus* LA-5 and *B. lactis* BB-12 for 8 weeks and improved liver function, led to a favorable change in serum lipid profiles (Nabavi et al. 2014). In addition to adults, similar effects have been seen in children. reatment with bifidobacteria, lactobacilli, and S. thermophilus strains for 4 months significantly reduced fatty liver severity (evaluated by ultrasound) by inhibiting hepatic fat accumulation and preventing NAFLD progression. The decrease in BMI in children was another finding of the mentioned study (Alisi et al. 2014). In summarizing these studies in a conducted meta-analysis by Ma et al. (2013) to evaluate the efficacy of probiotics in NAFLD patients, it was shown that supplementation with probiotics significantly reduced alanine aminotransferase, aspartate transaminase, total-cholesterol, HDL-C, TNF-α and HOMA-IR (insulin resistance index).

In addition to probiotics, frequent studies have pointed to the potential of prebiotics to treat NAFLD (Delzenne, Neyrinck, and Cani 2013; Daubioul et al. 2002; Daubioul et al. 2000; Fiordaliso et al. 1995). Prebiotics can influence the growth, activity and production of bacterial metabolites in the intestine (Gibson and Roberfroid 1995). These indigestible dietary ingredients can be fermented by beneficial bacteria such as bifidobacteria and lactobacilli in the intestine (Tannock et al. 2012). One of the most extensively studied and used prebiotics is fructooligosaccharide (FOS). Consuming prebiotics, including fructooligosaccharides, declines the growth of the pathogenic bacteria population, improves the growth and activity of favorable bacteria in the intestine, and in particular leads to bifidobacteria becoming the dominant species in the intestine (Gibson et al. 1995; Hofacre, Mathis, and Quiroz 2005). Probiotics have countless properties, of which some are referred to here. Increasing the content of bifidobacteria via prebiotics supplementation is associated with lower serum levels of endotoxins (Dewulf et al. 2012). Because the favorable microbiota modification due to prebiotics improves the integrity of the intestine and reduces the translocation of endotoxins (Roberfroid 2007), on the other hand, increased gut trophic hormone glucagon-like peptide-2 can reduce endotoxins by consolidating epithelial tight junctions (Cani et al. 2009). Finally, treatment with fructooligosaccharides by reducing inflammation and oxidative stress, improving glucose hemostasis and facilitating weight reduction can be effective in patients with NAFLD (Cani, Neyrinck, et al. 2007).

A recent four-armed clinical trial conducted by Behrouz et al. (2017) showed that supplementation with prebiotics and probiotics along with lifestyle modifications could lead to a satisfactory alteration in glycemic parameters and leptin levels. This intervention also managed to control the level of insulin and adipokines. In sum, probiotics and prebiotics can be considered adjunct therapy in the NAFLD's metabolic disturbances. Improving liver function, enhancing energy metabolism and reducing insulin resistance are mechanisms by which probiotics and prebiotics may treat NAFLD.

2.3.2. Cirrhosis

Since cirrhosis is an advanced liver disease that may quickly lead to a patient's death, limited research has been completed regarding the effects of bioactive compounds on cirrhosis. Elevated synthesis of fatty acids and triglyceride accumulation in hepatocytes is a key aspect in liver fibrosis that may result in cirrhosis (Shi, Fan, and Xiao 2010). Studies have shown that AMPK and SIRT3 play a central role in regulating the liver's metabolism of energy and fatty acids (Kincaid and Bossy-Wetzel 2013; Um et al. 2013; Zong et al. 2002). One of the bioactive compounds that can target these two regulators is curcumin. Curcumin is an active component of turmeric and polyphenol derived from the yellow pigment of this spice (Grant and Schneider 2000) that can prevent hepatic injury by up-regulating the expression of SIRT3 mRNA (Chenari, Safari, and Moradi 2017). In addition to liver disorders, the role of curcumin in animal models of intestinal and pancreatic diseases has also been revealed (Gukovsky et al. 2003). The curcumin's therapeutic benefits are based on antioxidant, anti-inflammatory and anti-cancer properties (Gupta and Rath 2014; M.-E. Wang, Chen, et al. 2012).

Although the fundamental mechanism of the inhibitory effect of curcumin on fibrosis and cirrhosis of the liver has not been completely determined, undoubtedly anti-inflammatory activity of curcumin plays a substantial role (Gupta and Rath 2014; Wang, Chen, et al. 2012). Accelerating LPS clearance and improving hepatic detoxification are other proposed mechanisms involved in curcumin effects. Furthermore, it has also been shown in animal studies that curcumin enhances the expression of LDL receptors (Cai et al. 2017). Curcumin exerts its anti-inflammatory effect by inhibiting NF-κB (Joe, Vijaykumar, and Lokesh 2004). On the other hand, curcumin can prevent collagen type I production by activating hepatic stellate cells (Kang et al. 2002). The sum of these properties enables curcumin to reduce the inflammation and fibrosis of the liver and prevent the progression of steatohepatitis to cirrhosis (Nanji et al. 2003; Leclercq et al. 2004).

Inflammation is a chief indicator of cirrhosis; hence it seems that carotenoids can treat disease by reducing cytotoxicity and inflammation. Oxidative stress also plays a central role in liver fibrosis by hepatic stellate cells activation, mediated via enhancing NF-κB activity. Certain antioxidants can prevent the increase of NF-κB activity and inhibit stellate cells activation in the liver (Lee et al. 1995). As part of the carotenoid family, lycopene and beta-carotene can strongly interact with ROS to reduce liver fibrosis (Stahl and Sies 1996). In reaction to polyunsaturated fatty acids of cell membranes, nucleotides and even proteins, free oxygen radicals can cause cell damage. The presence of antioxidants such as vitamins C and E, beta-carotene, selenium and antioxidant enzymes such as glutathione peroxidase and superoxide dismutase can mitigate tissues injury (Stahl and Sies 1996). Antioxidants also reduce lipid peroxidation, a known causative factor in cirrhosis (Bedossa et al. 1994). As a whole, inflammation and peroxidation of fatty acids are two important aspects in activating hepatic stellate cells, which supplementation of antioxidants can target them (Houglum et al. 1990; Kamimura et al. 1992).

The role of intestinal microbiota has been elucidated in the pathophysiology of cirrhosis (Minemura and Shimizu 2015; Quigley, Stanton, and Murphy 2013). According to Chen et al. (2011), a decrease in abundance of Bacteroidetes and Proteobacteria and Fusobacteria in feces of Chinese patients with cirrhosis is associated with hepatitis B virus (HBV) infection and alcohol drinking was noted. Also, a decrease of the genera *Eubacterium*, *Bacteroides* and *Alistipes* and an increase of *Veillonella*, *Clostridium*, *Streptococcus*, and *Prevotella* were observed (Qin et al. 2014).

In patients with cirrhosis, gastrointestinal dysfunction, malabsorption, pathogenic bacterial overgrowth, and portal hypertension are seen (Sadik et al. 2003). Intestinal microbial imbalance in cirrhotic patients is demonstrated as intestinal-bacterial interactions, increased intestinal permeability named leaky gut, translocation of LPS into the body and the development of infectious complications such as peritonitis and encephalopathy (Wiest, Lawson, and Geuking 2014; Bajaj et al. 2014; Minemura and Shimizu 2015; Garcovich et al. 2012). Endotoxins penetration into the body and endotoxemia occurrence in patients with cirrhosis is allied with systemic disturbances such as multiple organ failure, spontaneous bacterial peritonitis, hepatorenal syndrome and cirrhotic cardiomyopathy (Hiroshi Fukui 2015), which are account for substantial morbidity and mortality (Arvaniti et al. 2010).

Gut dysbiosis on the one hand and disturbances in the integrity of the intestinal barrier on the other hand, with translocation of bacteria and their by-products, lead to inflammatory reactions that promote cirrhosis and exacerbate the complications of the disease (Fukui 2017). Metabolic abnormalities in patients with cirrhosis may be due to alteration in gastrointestinal function, including malabsorption, changes in intestinal movements, and small intestinal bacterial overgrowth (SIBO) (Gunnarsdottir et al. 2003).

According to the stated statements, it is not unexpected that prescribing probiotics or prebiotics can help cure cirrhosis or improve symptoms and complications of the disease. Reduced quantity of pathogenic bacteria population in fecal bacteriological analysis in cirrhosis patients confirms the intestinal microbiome modification due to probiotic administration (Liu et al. 2004). Probiotics inhibit the growth of bacterial pathogens by intraluminal acidification, competing for nutrients and secreting antimicrobial compounds (Gorbach 2000; Liu et al. 2004). Based on the results of the previous studies, consumption of probiotics can be useful in preventing and treating hepatic cirrhosis complications. For example, the modification of intestinal flora due to probiotic administration is accompanied by lower ammonium production. In a study conducted in cirrhotic patients with minimal hepatic encephalopathy, it was found that administration of probiotics for one month led to a 50% reversal of encephalopathy (Liu et al. 2004).

The result of a meta-analysis of 5 trials conducted on the effect of probiotics on ammonium reduction, which three had positive results (Lunia et al. 2014; Agrawal et al. 2012; Pereg et al. 2011) and two studies failed to support this finding (Mittal et al. 2011; Bajaj et al. 2008), showed that although administration of probiotics could reduce ammonium, the difference was not statistically significant. However, the result of this study indicated that the probiotic was effective in preventing hepatic encephalopathy in patients with cirrhosis (Xu et al. 2014). However, the previous meta-analyses failed to confirm the efficiency of probiotics in treating hepatic encephalopathy. Therefore, further clinical trials are needed to censure probiotic therapy in cirrhotic patients (McGee et al. 2011).

Short-chain fatty acids include acetate, propionate and butyrate, which are by-products of intestinal bacteria, which in addition to providing fuel for enterocytes, also enhance the intestinal barrier function. Along with local effects on colonocytes, short-chain fatty acids also have systemic effects and inflammation reduction is one of them (Brahe, Astrup, and Larsen 2013). In other words, increasing the population of pathogenic bacteria in the intestine by producing the LPS and stimulating activity of nuclear factor-kappa B increases the production and secretion of pro-inflammatory cytokines; while reinforcement of favorable gut bacteria by prescribing probiotics or prebiotics can neutralize these effects by increasing the production of SCFAs (Usami et al. 2013; Ohira et al. 2011).

LPS-induced inflammatory status leads to hepatic injury, which is improved by oral therapy with tributyrin, a pro-drug of butyrate and a dairy food component (Miyoshi et al. 2011). Stimulating of tight junction and mucous production are also influenced by SCFAs (Brahe, Astrup, and Larsen 2013; Wang, Wang, et al. 2012). The increased formic acid concentration in the feces indicates increased production of SCFAs due to synbiotic supplementation (M. Usami et al. 2011).

Lactose-fermenting lactobacilli and bifidobacteria have been the most widely used probiotics in gut microflora modulation (Pinzone et al. 2012). Reduction of oxidative parameters in patients with alcoholic liver cirrhosis has been reported through supplementation of VSL#3, a probiotic preparation of eight strains of lactobacilli, bifidobacteria and *Streptococcus* (Loguercio et al. 2005). One of the interesting effects of probiotics is their role in restoring phagocytic neutrophil capacity, which appears to be due to a change in the secretion of IL-10 and the expression of Toll-Like Receptor4 (Stadlbauer et al. 2008). Verifying this assumption requires more research.

Consumption of prebiotics such as lactitol and lactulose, synthetic non-absorbable disaccharides, can produce acetic acid and lactic acid through bacterial fermentation in the large intestine. The resulting lower intraluminal pH prevents the growth of pathogenic bacteria and enhances the growth of

Lactobacillus (Pinzone et al. 2012; Riggio et al. 1990). Increasing *Lactobacillus* and bifidobacteria is associated with decreased plasma endotoxin levels in patients with chronic viral hepatic diseases (Chen et al. 2007). However, it has been stated that the effectiveness of synbiotic in reducing bacterial infections is higher than prebiotics (fiber) alone in liver transplant recipients (Rayes et al. 2005). Another interventional study disclosed that improvement in virus-induced cirrhosis in patients receiving synbiotic supplements (47% improvement) was much more significant than that in the prebiotic group (29%) and placebo (8%) (Liu et al. 2004). Synbiotic therapy is also associated with a significant reduction in endotoxemia (Stadlbauer et al. 2008).

Conclusion

This chapter reviewed the effect of different bioactive compounds on health status of gastrointestinal tract. It has been highlighted that bioactive compounds decrease the level of free radicals and improve the antioxidant defense system such as antioxidant enzymes that has a protective role against oxidative damage and also reduce the production of key inflammatory mediators. In addition, probiotics can strengthen the gut epithelial barrier and exhibit immunomodulatory effects which along with prebiotics can balance the gut microflora. However, further studies are required to specify the exact mechanisms of effects of bioactive compounds from different sources on human gastrointestinal health. Also, designing functional food products containing bioactive compounds and demonstration of their effect on gastrointestinal health can be an issue of future investigations for food manufacturers and researchers.

References

Abbassi, F., and Hani, K. 2012. "In vitro antibacterial and antifungal activities of Rhus tripartitum used as antidiarrhoeal in Tunisian folk medicine." *Natural Product Research* 26 (23): 2215-2218.

Abboud, P. A., Hake, P. W., Burroughs, T. J., Odoms, K., O'connor, M., Mangeshkar, P., Wong, H. R., and Zingarelli, B. 2008. "Therapeutic effect of epigallocatechin-3-gallate in a mouse model of colitis." *European Journal of Pharmacology* 579 (1-3): 411-417.

Abrams, S. A., Griffin, I. J., Hawthorne, K. M., Liang, L., Gunn, S. K., Darlington, G., and Ellis, K. J. 2005. "A combination of prebiotic short-and long-chain inulin-type fructans enhances calcium absorption and bone mineralization in young adolescents." *The American Journal of Clinical Nutrition* 82 (2): 471-476.

Abrams, S. A., Hawthorne, K. M., Aliu, O., Hicks, P. D., Chen, Z., and Griffin, I. J. 2007. "An inulin-type fructan enhances calcium absorption primarily via an effect on colonic absorption in humans." *The Journal of Nutrition* 137 (10): 2208-2212.

Abuelsaad, A. S., Mohamed, I., Allam, G., and Al-Solumani, A. A. 2013. "Antimicrobial and immunomodulating activities of hesperidin and ellagic acid against diarrheic Aeromonas hydrophila in a murine model." *Life Science* 93 (20): 714-22.

Adefegha, S. A. 2018. "Functional foods and nutraceuticals as dietary intervention in chronic diseases; novel perspectives for health promotion and disease prevention." *Journal of Dietary Supplements* 15 (6): 977-1009.

Adhikary, B., Yadav, S. K., Bandyopadhyay, S. K., and Chattopadhyay, S. 2011. "Role of the COX-independent pathways in the ulcer-healing action of epigallocatechin gallate." *Food & Function* 2 (6): 338-347.

Agrawal, A., Sharma, B. C., Sharma, P., and Sarin, S. K. 2012. "Secondary prophylaxis of hepatic encephalopathy in cirrhosis: an open-label, randomized controlled trial of lactulose, probiotics, and no therapy." *The American Journal of Gastroenterology* 107 (7): 1043.

Agréus, L., Svärdsudd, K., Nyrén, O., and Tibblin, G. 1995. "Irritable bowel syndrome and dyspepsia in the general population: overlap and lack of stability over time." *Gastroenterology* 109 (3): 671-680.

Aguirre, L., Portillo, M. P., Hijona, E., and Bujanda, L. 2014. "Effects of resveratrol and other polyphenols in hepatic steatosis." *World Journal of Gastroenterol* 20 (23): 7366-80.

Ahn, J., Cho, I., Kim, S., Kwon, D., and Ha, T. 2008. "Dietary resveratrol alters lipid metabolism-related gene expression of mice on an atherogenic diet." *Journal of Hepatology* 49 (6): 1019-1028.

Ait-Belgnaoui, A., Bradesi, S., Fioramonti, J., Theodorou, V., and Bueno, L. 2005. "Acute stress-induced hypersensitivity to colonic distension depends upon increase in paracellular permeability: role of myosin light chain kinase." *Pain* 113 (1-2): 141-147.

Aiyer, H. S., Li, Y., Liu, Q. H., Reuter, N., and Martin, R. C. 2011. "Dietary freeze-dried black raspberry's effect on cellular antioxidant status during reflux-induced esophagitis in rats." *Nutrition* 27 (2): 182-187.

Akhlaghi, M. 2016. "Non-alcoholic Fatty Liver Disease: Beneficial Effects of Flavonoids." *Phytotherapy Research* 30 (10): 1559-1571.

Alberdi, G., Rodríguez, V. M., Macarulla, M. T., Miranda, J., Churruca, I., and Portillo, M. P. 2013. "Hepatic lipid metabolic pathways modified by resveratrol in rats fed an obesogenic diet." *Nutrition* 29 (3): 562-567.

Aldoori, W. H., Giovannucci, E. L., Stampfer, M. J., Rimm, E. B., Wing, A. L., and Willett, W. C. 1997. "Prospective study of diet and the risk of duodenal ulcer in men." *American Journal of Epidemiology* 145 (1): 42-50.

Alisi, A., Bedogni, G., Baviera, G., Giorgio, V., Porro, E., Paris, C., Giammaria, P., Reali, L., Anania, F., and Nobili, V. 2014. "Randomised clinical trial: the beneficial effects of VSL# 3 in obese children with non-alcoholic steatohepatitis." *Alimentary Pharmacology & Therapeutics* 39 (11): 1276-1285.

Allen, S. J., Okoko, B., Martinez, E., Gregorio, G., and Dans, L. F. 2003. "Probiotics for treating infectious diarrhoea." *Cochrane Database Syst Rev* 4.

Aller, R., De Luis, D., Izaola, O., Conde, R., Gonzalez Sagrado, M., Primo, D., De La Fuente, B., and Gonzalez, J. 2011. "Effect of a probiotic on liver aminotransferases in nonalcoholic fatty liver disease patients: a double blind randomized clinical trial." *European Review for Medical Pharmacological Sciences* 15 (9): 1090-5.

Amre, D. K., D'souza, S., Morgan, K., Seidman, G., Lambrette, P., Grimard, G., Israel, D., Mack, D., Ghadirian, P., and Deslandres, C. 2007. "Imbalances in dietary consumption of fatty acids, vegetables, and fruits are associated with risk for Crohn's disease in children." *The American Journal of Gastroenterology* 102 (9): 2016.

Anderson, J. W., Baird, P., Davis, R. H., Jr., Ferreri, S., Knudtson, M., Koraym, A., Waters, V., and Williams, C. L. 2009. "Health benefits of dietary fiber." *Nutrition Review* 67 (4): 188-205.

Andersson, A. F., Lindberg, M., Jakobsson, H., Bäckhed, F., Nyrén, P., and Engstrand, L. 2008. "Comparative analysis of human gut microbiota by barcoded pyrosequencing." *PloS one* 3 (7): e2836.

Andújar, I., Recio, M. C., Giner, R. M., Cienfuegos-Jovellanos, E., Laghi, S., Muguerza, B. a., and Ríos, J. L. 2011. "Inhibition of ulcerative colitis in mice after oral administration of a polyphenol-enriched cocoa extract is mediated by the inhibition of STAT1 and STAT3 phosphorylation in colon cells." *Journal of Agricultural and Food Chemistry* 59 (12): 6474-6483.

Arvaniti, V., D'Amico, G., Fede, G., Manousou, P., Tsochatzis, E., Pleguezuelo, M., and Burroughs, A. K. 2010. "Infections in patients with cirrhosis increase mortality four-fold and should be used in determining prognosis." *Gastroenterology* 139 (4): 1246-1256. e5.

Ashok, P. K., and Upadhyaya, K. 2012. "Tannins are astringent." *Journal of Pharmacognosy and Phytochemistry* 1 (3).

Attaluri, A., Jackson, M., Valestin, J., and Rao, S. S. 2010. "Methanogenic flora is associated with altered colonic transit but not stool characteristics in constipation without IBS." *The American Journal of Gastroenterology* 105 (6): 1407-1411.

Aubrecht, E., Horacsek, M., Gelencser, E., and Dworschak, E. 1998. "Investigation of prolamin content of cereals and different plant seeds." *Acta Alimentaria* 27 (2): 119-125.

Awad, A. B., Toczek, J., and Fink, C. S. 2004. "Phytosterols decrease prostaglandin release in cultured P388D 1/MAB macrophages." *Prostaglandins, Leukotrienes and Essential Fatty Acids* 70 (6): 511-520.

Azuma, T., Shigeshiro, M., Kodama, M., Tanabe, S., and Suzuki, T. 2013. "Supplemental Naringenin Prevents Intestinal Barrier Defects and Inflammation in Colitic Mice–3." *The Journal of Nutrition* 143 (6): 827-834.

Babu, A., Pon, V., and Liu, D. 2008. "Green tea catechins and cardiovascular health: an update." *Current Medicinal Chemistry* 15 (18): 1840-1850.

Bäckhed, F., Ley, R. E., Sonnenburg, J. L., Peterson, D. A., and Gordon, J. I. 2005. "Host-bacterial mutualism in the human intestine." *Science* 307 (5717): 1915-1920.

Bahcecioglu, I. H., Kuzu, N., Metin, K., Ozercan, I. H., Ustündag, B., Sahin, K., and Kucuk, O. 2010. "Lycopene prevents development of steatohepatitis in experimental nonalcoholic steatohepatitis model induced by high-fat diet." *Veterinary Medicine International* 2010.

Bajaj, J. S., Heuman, D. M., Hylemon, P. B., Sanyal, A. J., Puri, P., Sterling, R. K., Luketic, V., Stravitz, R. T., Siddiqui, M. S., and Fuchs, M. 2014. "Randomised clinical trial: Lactobacillus GG modulates gut microbiome, metabolome and endotoxemia in patients with cirrhosis." *Alimentary Pharmacology & Therapeutics* 39 (10): 1113-1125.

Bajaj, J. S., Saeian, K., Christensen, K. M., Hafeezullah, M., Varma, R. R., Franco, J., Pleuss, J. A., Krakower, G., Hoffmann, R. G., and Binion, D. G. 2008. "Probiotic yogurt for the treatment of minimal hepatic encephalopathy." *The American Journal of Gastroenterology* 103 (7): 1707.

Banerjee, D., Bhattacharya, S., Bandyopadhyay, S. K., and Chattopadhyay, S. 2008. "Biochemical mechanism of healing activity of the natural phenolic, allylpyrocatechol against indomethacin-induced gastric ulceration in mice." *Digestive Diseases and Sciences* 53 (11): 2868.

Baur, J. A., Pearson, K. J., Price, N. L., Jamieson, H. A., Lerin, C., Kalra, A., Prabhu, V. V., Allard, J. S., Lopez-Lluch, G., and Lewis, K. 2006. "Resveratrol improves health and survival of mice on a high-calorie diet." *Nature* 444 (7117): 337-342.

Bedossa, P., Houglum, K., Trautwein, C., Holstege, A., and Chojkier, M. 1994. "Stimulation of collagen α1 (I) gene expression is associated with lipid peroxidation in hepatocellular injury: a link to tissue fibrosis?" *Hepatology* 19 (5): 1262-1271.

Behrouz, V., Jazayeri, S., Aryaeian, N., Zahedi, M. J., and Hosseini, F. 2017. "Effects of Probiotic and Prebiotic Supplementation on Leptin, Adiponectin, and Glycemic Parameters in Non-alcoholic Fatty Liver Disease: A Randomized Clinical Trial." *Middle East Journal of Digestive Diseases* 9 (3): 150-157.

Benjamin, J. L., Hedin, C. R., Koutsoumpas, A., Ng, S. C., McCarthy, N. E., Hart, A. L., Kamm, M. A., Sanderson, J. D., Knight, S. C., and Forbes, A. 2011. "Randomised, double-blind, placebo-controlled trial of fructo-oligosaccharides in active Crohn's disease." *Gut* 60 (7): 923-929.

Bernal, C., Martín-Pozuelo, G., Lozano, A. B., Sevilla, Á., García-Alonso, J., Canovas, M., and Periago, M. J. 2013. "Lipid biomarkers and metabolic effects of lycopene from tomato juice on liver of rats with induced hepatic steatosis." *The Journal of Nutritional Biochemistry* 24 (11): 1870-1881.

Bhathena, J., Martoni, C., Kulamarva, A., Tomaro-Duchesneau, C., Malhotra, M., Paul, A., Urbanska, A. M., and Prakash, S. 2013. "Oral probiotic microcapsule formulation ameliorates non-alcoholic fatty liver disease in Bio F1B Golden Syrian hamsters." *PLoS One* 8 (3): e58394.

Bibiloni, R., Fedorak, R. N., Tannock, G. W., Madsen, K. L., Gionchetti, P., Campieri, M., De Simone, C., and Sartor, R. B. 2005. "VSL# 3 probiotic-mixture induces remission in patients with active ulcerative colitis." *The American Journal of Gastroenterology* 100 (7): 1539.

Biedermann, L., Mwinyi, J., Scharl, M., Frei, P., Zeitz, J., Kullak-Ublick, G. A., Vavricka, S. R., Fried, M., Weber, A., and Humpf, H.-U. 2013. "Bilberry ingestion improves disease activity in mild to moderate ulcerative colitis—An open pilot study." *Journal of Crohn's and Colitis* 7 (4): 271-279.

Biesalski, H.-K., Dragsted, L. O., Elmadfa, I., Grossklaus, R., Müller, M., Schrenk, D., Walter, P., and Weber, P. 2009. "Bioactive compounds: Definition and assessment of activity." *Nutrition* 25 (11-12): 1202-1205.

Bijkerk, C., De Wit, N., Muris, J., Whorwell, P., Knottnerus, J., and Hoes, A. 2009. "Soluble or insoluble fibre in irritable bowel syndrome in primary care? Randomised placebo controlled trial." *Bmj* 339: b3154.

Birk, Y. 1985. "The Bowman-Birk inhibitor. Trypsin-and chymotrypsin- inhibitor from soybeans." *Chemical Biology & Drug Design* 25 (2): 113-131.

Bischoff, S. C. 2008. "Quercetin: potentials in the prevention and therapy of disease." *Current Opinion in Clinical Nutrition & Metabolic Care* 11 (6): 733-740.

Bomba, A., Nemcova, R., Gancarcikova, S., Herich, R., Guba, P., and Mudronova, D. 2002. "Improvement of the probiotic effect of micro-organisms by their combination with maltodextrins, fructo-oligosaccharides and polyunsaturated fatty acids." *British Journal of Nutrition* 88 (S1): S95-S99.

Bose, M., Lambert, J. D., Ju, J., Reuhl, K. R., Shapses, S. A., and Yang, C. S. 2008. "The major green tea polyphenol,(-)-epigallocatechin-3-gallate, inhibits obesity, metabolic syndrome, and fatty liver disease in high-fat–fed mice." *The Journal of Nutrition* 138 (9): 1677-1683.

Bradford, P. G. 2013. "Curcumin and obesity." *Biofactors* 39 (1): 78-87.

Brahe, L. K., Astrup, A., and Larsen, L. H. 2013. "Is butyrate the link between diet, intestinal microbiota and obesity-related metabolic diseases?" *Obesity Reviews* 14 (12): 950-959.

Brandt, L. J., Prather, C. M., Quigley, E. M., Schiller, L. R., Schoenfeld, P., and Talley, N. J. 2005. "Systematic review on the management of chronic constipation in North America." *American Journal of Gastroenterology* 100 (s1): S5-S21.

Braniste, V., Leveque, M., Buisson-Brenac, C., Bueno, L., Fioramonti, J., and Houdeau, E. 2009. "Oestradiol decreases colonic permeability through oestrogen receptor β-mediated up-regulation of occludin and junctional adhesion molecule-A in epithelial cells." *The Journal of Physiology* 587 (13): 3317-3328.

Bruno, R. S., Dugan, C. E., Smyth, J. A., DiNatale, D. A., and Koo, S. I. 2008. "Green tea extract protects leptin-deficient, spontaneously obese mice from hepatic steatosis and injury." *The Journal of Nutrition* 138 (2): 323-331.

Bujanda, L., Hijona, E., Larzabal, M., Beraza, M., Aldazabal, P., García-Urkia, N., Sarasqueta, C., Cosme, A., Irastorza, B., and González, A. 2008. "Resveratrol inhibits nonalcoholic fatty liver disease in rats." *BMC Gastroenterology* 8 (1): 40.

Bukhari, I., Shah, A., Khan, R., Meo, S., Khan, A., and Gilani, A. 2013. "Gut modulator effects of Conyza bonariensis explain its traditional use in constipation and diarrhea." *European Review for Medical and Pharmacological Sciences* 17 (4): 552.

Cai, Y., Lu, D., Zou, Y., Zhou, C., Liu, H., Tu, C., Li, F., Liu, L., and Zhang, S. 2017. "Curcumin Protects Against Intestinal Origin Endotoxemia in Rat Liver Cirrhosis by Targeting PCSK9." *Journal of Food Science* 82 (3): 772-780.

Calatayud, S., García-Zaragozá, E., Hernández, C., Quintana, E., Felipo, V., Esplugues, J. V., and Barrachina, M. D. 2002. "Downregulation of nNOS and synthesis of PGs associated with endotoxin-induced delay in gastric emptying." *American Journal of Physiology-Gastrointestinal and Liver Physiology* 283 (6): G1360-G1367.

Calder, P. C. 2011. "Fatty acids and inflammation: the cutting edge between food and pharma." *European Journal of Pharmacology* 668: S50-S58.

Calder, P. C., Albers, R., Antoine, J.-M., Blum, S., Bourdet-Sicard, R., Ferns, G., Folkerts, G., Friedmann, P., Frost, G., and Guarner, F. 2009. "Inflammatory disease processes and interactions with nutrition." *British Journal of Nutrition* 101 (S1): 1-45.

Camilleri, M. 2006. "Is there a role for probiotics in irritable bowel syndrome?" *Digestive and Liver Disease* 38: S266-S269.

Cani, P. D., Amar, J., Iglesias, M. A., Poggi, M., Knauf, C., Bastelica, D., Neyrinck, A. M., Fava, F., Tuohy, K. M., and Chabo, C. 2007. "Metabolic endotoxemia initiates obesity and insulin resistance." *Diabetes* 56 (7): 1761-1772.

Cani, P. D., Neyrinck, A., Fava, F., Knauf, C., Burcelin, R., Tuohy, K., Gibson, G., and Delzenne, N. 2007. "Selective increases of bifidobacteria in gut microflora improve high-fat-diet-induced diabetes in mice through a mechanism associated with endotoxaemia." *Diabetologia* 50 (11): 2374-2383.

Cani, P. D., Possemiers, S., Van de Wiele, T., Guiot, Y., Everard, A., Rottier, O., Geurts, L., Naslain, D., Neyrinck, A., and Lambert, D. M. 2009. "Changes in gut microbiota control inflammation in obese mice through a mechanism involving GLP-2-driven improvement of gut permeability." *Gut* 58 (8): 1091-1103.

Cano, P. G., Santacruz, A., Trejo, F. M., and Sanz, Y. 2013. "Bifidobacterium CECT 7765 improves metabolic and immunological alterations associated with obesity in high-fat diet-fed mice." *Obesity* 21 (11): 2310-2321.

Cao, Y., Wang, C., Liu, J., Liu, Z. M., Ling, W. H., and Chen, Y. M. 2015. "Greater serum carotenoid levels associated with lower prevalence of nonalcoholic fatty liver disease in Chinese adults." *Scientific Reports* 5: 12951.

Capili, B., Anastasi, J. K., and Chang, M. 2016. "Addressing the role of food in irritable bowel syndrome symptom management." *The Journal for Nurse Practitioners* 12 (5): 324-329.

Capriles, V. D., Martini, L. A., and Arêas, J. A. G. 2009. "Metabolic osteopathy in celiac disease: importance of a gluten-free diet." *Nutrition Reviews* 67 (10): 599-606.

Cencic, A., and Chingwaru, W. 2010. "The role of functional foods, nutraceuticals, and food supplements in intestinal health." *Nutrients* 2 (6): 611-625.

Chachay, V. S., Macdonald, G. A., Martin, J. H., Whitehead, J. P., O'Moore–Sullivan, T. M., Lee, P., Franklin, M., Klein, K., Taylor, P. J., and Ferguson, M. 2014. "Resveratrol does not benefit patients with nonalcoholic fatty liver disease." *Clinical Gastroenterology and Hepatology* 12 (12): 2092-2103. e6.

Chandran, P., Satthaporn, S., Robins, A., and Eremin, O. 2003. "Inflammatory bowel disease: dysfunction of GALT and gut bacterial flora (II)." *The Surgeon* 1 (3): 125-136.

Chattopadhyay, I., Bandyopadhyay, U., Biswas, K., Maity, P., and Banerjee, R. K. 2006. "Indomethacin inactivates gastric peroxidase to induce reactive-oxygen-mediated gastric mucosal injury and curcumin protects it by preventing peroxidase inactivation and scavenging reactive oxygen." *Free Radical Biology and Medicine* 40 (8): 1397-1408.

Chaves, G. V., Pereira, S. E., Saboya, C. J., and Ramalho, A. 2008. "Non-alcoholic fatty liver disease and its relationship with the nutritional status of vitamin A in individuals with class III obesity." *Obesity Surgery* 18 (4): 378-385.

Chen, C., Li, L., Wu, Z., Chen, H., and Fu, S. 2007. "Effects of lactitol on intestinal microflora and plasma endotoxin in patients with chronic viral hepatitis." *Journal of Infection* 54 (1): 98-102.

Chen, J.-H., Tipoe, G. L., Liong, E. C., So, H. S., Leung, K.-M., Tom, W.-M., Fung, P. C., and Nanji, A. A. 2004. "Green tea polyphenols prevent toxin-induced hepatotoxicity in mice by down-regulating inducible nitric oxide–derived prooxidants." *The American Journal of Clinical Nutrition* 80 (3): 742-751.

Chen, S., Zhao, X., Ran, L., Wan, J., Wang, X., Qin, Y., Shu, F., Gao, Y., Yuan, L., and Zhang, Q. 2015. "Resveratrol improves insulin resistance, glucose and lipid metabolism in patients with non-alcoholic fatty liver disease: a randomized controlled trial." *Digestive and Liver Disease* 47 (3): 226-232.

Chen, Y., Yang, F., Lu, H., Wang, B., Chen, Y., Lei, D., Wang, Y., Zhu, B., and Li, L. 2011. "Characterization of fecal microbial communities in patients with liver cirrhosis." *Hepatology* 54 (2): 562-572.

Chenari, S., Safari, F., and Moradi, A. 2017. "Curcumin enhances liver SIRT3 expression in the rat model of cirrhosis." *Iran J Basic Med Sci* 20 (12): 1306-1311.

Chirumbolo, S. 2010. "The role of quercetin, flavonols and flavones in modulating inflammatory cell function." *Inflammation & Allergy-Drug Targets* 9 (4): 263-285.

Chuengsamarn, S., Rattanamongkolgul, S., Luechapudiporn, R., Phisalaphong, C., and Jirawatnotai, S. 2012. "Curcumin extract for prevention of type 2 diabetes." *Diabetes Care* 35 (11): 2121-2127.

Chung, M. Y., Mah, E., Masterjohn, C., Noh, S. K., Park, H. J., Clark, R. M., Park, Y. K., Lee, J. Y., and Bruno, R. S. 2015. "Green Tea Lowers Hepatic COX-2 and Prostaglandin E2 in Rats with Dietary Fat-Induced Nonalcoholic Steatohepatitis." *Journal of Medicinal Food* 18 (6): 648-55.

Chutkan, R., Fahey, G., Wright, W. L., and McRorie, J. 2012. "Viscous versus nonviscous soluble fiber supplements: Mechanisms and evidence for fiber-specific health benefits." *Journal of the American Association of Nurse Practitioners* 24 (8): 476-487.

Cimperman, L., Bayless, G., Best, K., Diligente, A., Mordarski, B., Oster, M., Smith, M., Vatakis, F., Wiese, D., and Steiber, A. 2011. "A randomized, double-blind, placebo-controlled pilot study of Lactobacillus reuteri ATCC 55730 for the prevention of antibiotic-associated diarrhea in hospitalized adults." *Journal of Clinical Gastroenterology* 45 (9): 785-789.

Coconnier, M.-H., Lievin, V., Hemery, E., and Servin, A. L. 1998. "Antagonistic activity againstHelicobacter infection in vitro and in vivo by the humanLactobacillus acidophilus strain LB." *Applied and Environmental Microbiology* 64 (11): 4573-4580.

Collado, M. C., Donat, E., Ribes-Koninckx, C., Calabuig, M., and Sanz, Y. 2008. "Specific duodenal and faecal bacterial groups are associated with paediatric celiac disease." *Journal of Clinical Pathology*.

Commission, C. G. 2001. "Nutritional profile of No. 1 Canada Western flaxseed and of yellow flaxseed samples." *Can. Grain Comm., Winnipeg, Manitoba.*

Compare, D., Coccoli, P., Rocco, A., Nardone, O., De Maria, S., Cartenì, M., and Nardone, G. 2012. "Gut–liver axis: the impact of gut microbiota on non- alcoholic fatty liver disease." *Nutrition, Metabolism and Cardiovascular Diseases* 22 (6): 471-476.

Conquer, J., Maiani, G., Azzini, E., Raguzzini, A., and Holub, B. 1998. "Supplementation with quercetin markedly increases plasma quercetin concentration without effect on selected risk factors for heart disease in healthy subjects." *The Journal of Nutrition* 128 (3): 593-597.

Constantin, R. P., Nascimento, G. S. d., Constantin, R. P., Salgueiro, C. L., Bracht, A., Ishii-Iwamoto, E. L., Yamamoto, N. S., and Constantin, J. 2013. "Citrus flavanones affect hepatic fatty acid oxidation in rats by acting as prooxidant agents." *BioMed Research International* 2013.

Conteas, C. N., Panossian, A. M., Tran, T. T., and Singh, H. M. 2009. "Treatment of HIV-associated diarrhea with curcumin." *Digestive Diseases and Sciences* 54 (10): 2188-91.

Cowan, T. E., Palmnäs, M. S., Yang, J., Bomhof, M. R., Ardell, K. L., Reimer, R. A., Vogel, H. J., and Shearer, J. 2014. "Chronic coffee consumption in the diet-induced obese rat: impact on gut microbiota and serum metabolomics." *The Journal of Nutritional Biochemistry* 25 (4): 489-495.

Crespillo, A., Alonso, M., Vida, M., Pavón, F., Serrano, A., Rivera, P., Romero-Zerbo, Y., Fernández-Llebrez, P., Martínez, A., and Pérez-Valero, V. 2011. "Reduction of body weight, liver steatosis and expression of stearoyl-CoA desaturase 1 by the isoflavone daidzein in diet-induced obesity." *British Journal of Pharmacology* 164 (7): 1899-1915.

Crespo, I., García-Mediavilla, M. V., Almar, M., González, P., Tuñón, M. J., Sánchez-Campos, S., and González-Gallego, J. 2008. "Differential effects of dietary flavonoids on reactive oxygen and nitrogen species generation and changes in antioxidant enzyme expression induced by proinflammatory cytokines in Chang Liver cells." *Food and Chemical Toxicology* 46 (5): 1555-1569.

Cuevas, M. J., Tieppo, J., Marroni, N. P., Tuñón, M. J., and González-Gallego, J. 2011. "Suppression of amphiregulin/epidermal growth factor receptor signals contributes to the protective effects of quercetin in cirrhotic rats." *The Journal of Nutrition* 141 (7): 1299-1305.

Cummings, J., and Macfarlane, G. 2002. "Gastrointestinal effects of prebiotics." *British Journal of Nutrition* 87 (S2): S145-S151.

D'souza, A. L., Rajkumar, C., Cooke, J., and Bulpitt, C. J. 2002. "Probiotics in prevention of antibiotic associated diarrhoea: meta-analysis." *Bmj* 324 (7350): 1361.

Daubioul, C., Rousseau, N., Demeure, R., Gallez, B., Taper, H., Declerck, B., and Delzenne, N. 2002. "Dietary fructans, but not cellulose, decrease triglyceride accumulation in the liver of obese Zucker fa/fa rats." *The Journal of Nutrition* 132 (5): 967-973.

Daubioul, C. A., Taper, H. S., Laurent, D., and Delzenne, N. M. 2000. "Dietary oligofructose lessens hepatic steatosis, but does not prevent hypertriglyceridemia in obese Zucker rats." *The Journal of Nutrition* 130 (5): 1314-1319.

De Keersmaecker, S. C., Verhoeven, T. L., Desair, J., Marchal, K., Vanderleyden, J., and Nagy, I. 2006. "Strong antimicrobial activity of Lactobacillus rhamnosus GG against

Salmonella typhimurium is due to accumulation of lactic acid." *FEMS Microbiology Letters* 259 (1): 89-96.

De Palma, G., Nadal, I., Medina, M., Donat, E., Ribes-Koninckx, C., Calabuig, M., and Sanz, Y. 2010. "Intestinal dysbiosis and reduced immunoglobulin-coated bacteria associated with coeliac disease in children." *BMC Microbiology* 10: 63.

De, R., Kundu, P., Swarnakar, S., Ramamurthy, T., Chowdhury, A., Nair, G. B., and Mukhopadhyay, A. K. 2009. "Antimicrobial activity of curcumin against Helicobacter pylori isolates from India and during infections in mice." *Antimicrobial Agents and Chemotherapy* 53 (4): 1592-1597.

De Stefano, D., Maiuri, M. C., Simeon, V., Grassia, G., Soscia, A., Cinelli, M. P., and Carnuccio, R. 2007. "Lycopene, quercetin and tyrosol prevent macrophage activation induced by gliadin and IFN-γ." *European Journal of Pharmacology* 566 (1): 192-199.

de Vrese, M., and Marteau, P. R. 2007. "Probiotics and prebiotics: effects on diarrhea." *J Nutr* 137 (3 Suppl 2): 803s-11s.

Delzenne, N. M., Neyrinck, A. M., and Cani, P. D. 2013. "Gut microbiota and metabolic disorders: how prebiotic can work?" *British Journal of Nutrition* 109 (S2): S81-S85.

Dewulf, E. M., Cani, P. D., Claus, S. P., Fuentes, S., Puylaert, P. G., Neyrinck, A. M., Bindels, L. B., de Vos, W. M., Gibson, G. R., and Thissen, J.-P. 2012. "Insight into the prebiotic concept: lessons from an exploratory, double blind intervention study with inulin-type fructans in obese women." *Gut* 62 (8): 1112-1121.

Di Carlo, G., Mascolo, N., Izzo, A. A., and Capasso, F. 1999. "Flavonoids: old and new aspects of a class of natural therapeutic drugs." *Life Sciences* 65 (4): 337-353.

Didari, T., Mozaffari, S., Nikfar, S., and Abdollahi, M. 2015. "Effectiveness of probiotics in irritable bowel syndrome: Updated systematic review with meta-analysis." *World Journal of Gastroenterology* 21 (10): 3072-3084.

Distrutti, E., Monaldi, L., Ricci, P., and Fiorucci, S. 2016. "Gut microbiota role in irritable bowel syndrome: New therapeutic strategies." *World Journal of Gastroenterology* 22 (7): 2219-2241.

Do, G. M., Jung, U. J., Park, H. J., Kwon, E. Y., Jeon, S. M., McGregor, R. A., and Choi, M. S. 2012. "Resveratrol ameliorates diabetes-related metabolic changes via activation of AMP-activated protein kinase and its downstream targets in db/db mice." *Molecular Nutrition & Food Research* 56 (8): 1282-1291.

Dolin, B. 2009. "Effects of a proprietary Bacillus coagulans preparation on symptoms of diarrhea-predominant irritable bowel syndrome." *Methods and Findings in Experimental and Clinical Pharmacology* 31 (10): 655-659.

Dover, A., Patel, N., and Park, K. T. 2015. "Rapid cessation of acute diarrhea using a novel solution of bioactive polyphenols: a randomized trial in Nicaraguan children." *PeerJ* 3: e969.

Dryden, G. W., Lam, A., Beatty, K., Qazzaz, H. H., and McClain, C. J. 2013. "A Pilot Study to Evaluate the Safety and Efficacy of an Oral Dose of (−)-Epigallocatechin-3-Gallate–Rich Polyphenon E in Patients with Mild to Moderate Ulcerative Colitis." *Inflammatory Bowel Diseases* 19 (9): 1904-1912.

Duke, J., Bogenschutz-Godwin, M., duCellier, M., and Duke, P. 2002. Catalog of herbs (A–Z): flax, Handbook of Medicinal Herbs. CRC press, Boca Raton, Florida.

Duseja, A., and Chawla, Y. K. 2014. "Obesity and NAFLD." *Clinics in Liver Disease* 18 (1): 59-71.

Dutta, S. K., Agrawal, K., Girotra, M., Fleisher, A. S., Motevalli, M., Mah'moud, M. A., and Nair, P. P. 2012. "Barrett's esophagus and beta-carotene therapy: symptomatic improvement in GERD and enhanced HSP70 expression in esophageal mucosa." *Asian Pacific Journal of Cancer Prevention* 13 (12): 6011-6.

El-Salhy, M. 2012. *Irritable bowel syndrome: Diagnosis, pathogenesis, and treatment options*. Nova Biomedical, Nova Science Publishers, Incorporated.

El-Salhy, M., and Gundersen, D. 2015. "Diet in irritable bowel syndrome." *Nutrition Journal* 14: 36.

El-Serag, H., Satia, J., and Rabeneck, L. 2005. "Dietary intake and the risk of gastro-oesophageal reflux disease: a cross sectional study in volunteers." *Gut* 54 (1): 11-17.

Elitsur, Y., and Yahav, J. 2005. "Helicobacter pylori infection in pediatrics." *Helicobacter* 10 (s1): 47-53.

Elmaliklis, I.-N., Liveri, A., Ntelis, B., Paraskeva, K., Goulis, I., and Koutelidakis, A. E. 2019. "Increased functional foods' consumption and mediterranean diet adherence may have a protective effect in the appearance of gastrointestinal diseases: A case–control study." *Medicines* 6 (2): 50.

Elshazly, S. M., and Mahmoud, A. A. 2014. "Antifibrotic activity of hesperidin against dimethylnitrosamine-induced liver fibrosis in rats." *Naunyn-Schmiedeberg's Archives of Pharmacology* 387 (6): 559-567.

Enck, P., Zimmermann, K., Menke, G., Müller-lissner, S., Martens, U., and Klosterhalfen, S. 2008. "A mixture of Escherichia coli (DSM 17252) and Enterococcus faecalis (DSM 16440) for treatment of the irritable bowel syndrome–a randomized controlled trial with primary care physicians." *Neurogastroenterology & Motility* 20 (10): 1103-1109.

Erhardt, A., Stahl, W., Sies, H., Lirussi, F., Donner, A., and Häussinger, D. 2011. "Plasma levels of vitamin E and carotenoids are decreased in patients with Nonalcoholic Steatohepatitis (NASH)." *European Journal of Medical Research* 16 (2): 76.

Eslamparast, T., Poustchi, H., Zamani, F., Sharafkhah, M., Malekzadeh, R., and Hekmatdoost, A. 2014. "Synbiotic supplementation in nonalcoholic fatty liver disease: a randomized, double-blind, placebo-controlled pilot study." *The American Journal of Clinical Nutrition* 99 (3): 535-542.

Espley, R. V., Butts, C. A., Laing, W. A., Martell, S., Smith, H., McGhie, T. K., Zhang, J., Paturi, G., Hedderley, D., and Bovy, A. 2013. "Dietary Flavonoids from Modified Apple Reduce Inflammation Markers and Modulate Gut Microbiota in Mice–3." *The Journal of Nutrition* 144 (2): 146-154.

Eswaran, S., Muir, J., and Chey, W. D. 2013. "Fiber and functional gastrointestinal disorders." *The American Journal of Gastroenterology* 108 (5): 718.

Eswaran, S., Tack, J., and Chey, W. D. 2011. "Food: the forgotten factor in the irritable bowel syndrome." *Gastroenterology Clinics* 40 (1): 141-162.

Etxeberria, U., Fernández-Quintela, A., Milagro, F. I., Aguirre, L., Martínez, J. A., and Portillo, M. P. 2013. "Impact of polyphenols and polyphenol-rich dietary sources on gut microbiota composition." *Journal of Agricultural and Food Chemistry* 61 (40): 9517-9533.

Faghihzadeh, F., Adibi, P., and Hekmatdoost, A. 2015. "The effects of resveratrol supplementation on cardiovascular risk factors in patients with non-alcoholic fatty liver disease: a randomised, double-blind, placebo-controlled study." *British Journal of Nutrition* 114 (5): 796-803.

Faghihzadeh, F., Adibi, P., Rafiei, R., and Hekmatdoost, A. 2014. "Resveratrol supplementation improves inflammatory biomarkers in patients with nonalcoholic fatty liver disease." *Nutrition Research* 34 (10): 837-843.

Fan, Y.P., Chakder, S., Gao, F., and Rattan, S. 2001. "Inducible and neuronal nitric oxide synthase involvement in lipopolysaccharide-induced sphincteric dysfunction." *American Journal of Physiology-Gastrointestinal and Liver Physiology* 280 (1): G32-G42.

Farah, A., Monteiro, M., Donangelo, C. M., and Lafay, S. 2008. "Chlorogenic acids from green coffee extract are highly bioavailable in humans." *The Journal of Nutrition* 138 (12): 2309-2315.

Farzaei, M.H., Abdollahi, M., and Rahimi, R. 2015. "Role of dietary polyphenols in the management of peptic ulcer." *World Journal of Gastroenterology* 21 (21): 6499-517.

Farzaei, M.H., Rahimi, R., and Abdollahi, M. 2015. "The role of dietary polyphenols in the management of inflammatory bowel disease." *Current Pharmaceutical Biotechnology* 16 (3): 196-210.

Fernandez-Banares, F., Hinojosa, J., Sanchez-Lombrana, J., Navarro, E., Martınez-Salmerón, J., Garcıa-Pugés, A., Gonzalez-Huix, F., Riera, J., Gonzalez-Lara, V., and Domınguez-Abascal, F. 1999. "Randomized clinical trial of Plantago ovata seeds (dietary fiber) as compared with mesalamine in maintaining remission in ulcerative colitis." *The American Journal of Gastroenterology* 94 (2): 427-433.

Ferretti, G., Bacchetti, T., Masciangelo, S., and Saturni, L. 2012. "Celiac disease, inflammation and oxidative damage: a nutrigenetic approach." *Nutrients* 4 (4): 243-57.

Fiordaliso, M., Kok, N., Desager, J.-P., Goethals, F., Deboyser, D., Roberfroid, M., and Delzenne, N. 1995. "Dietary oligofructose lowers triglycerides, phospholipids and cholesterol in serum and very low density lipoproteins of rats." *Lipids* 30 (2): 163-167.

Floch, M. H., Walker, W. A., Madsen, K., Sanders, M. E., Macfarlane, G. T., Flint, H. J., Dieleman, L. A., Ringel, Y., Guandalini, S., and Kelly, C. P. 2011. "Recommendations for probiotic use—2011 update." *Journal of Clinical Gastroenterology* 45: S168-S171.

Ford, A. C., Moayyedi, P., Lacy, B. E., Lembo, A. J., Saito, Y. A., Schiller, L. R., Soffer, E. E., Spiegel, B. M., and Quigley, E. M. 2014. "American College of Gastroenterology monograph on the management of irritable bowel syndrome and chronic idiopathic constipation." *The American Journal of Gastroenterology* 109 (S1): S2.

Ford, A. C., Quigley, E. M., Lacy, B. E., Lembo, A. J., Saito, Y. A., Schiller, L. R., Soffer, E. E., Spiegel, B. M., and Moayyedi, P. 2014. "Efficacy of prebiotics, probiotics, and synbiotics in irritable bowel syndrome and chronic idiopathic constipation: systematic review and meta-analysis." *The American Journal of Gastroenterology* 109 (10): 1547.

Ford, A. C., Talley, N. J., Spiegel, B. M., Foxx-Orenstein, A. E., Schiller, L., Quigley, E. M., and Moayyedi, P. 2008. "Effect of fibre, antispasmodics, and peppermint oil in the treatment of irritable bowel syndrome: systematic review and meta-analysis." *Bmj* 337: a2313.

Foryst-Ludwig, A., Neumann, M., Schneider-Brachert, W., and Naumann, M. 2004. "Curcumin blocks NF-κB and the motogenic response in Helicobacter pylori-infected epithelial cells." *Biochemical and Biophysical Research Communications* 316 (4): 1065-1072.

Francis, C., and Whorwell, P. 1994. "Bran and irritable bowel syndrome: time for reappraisal." *The Lancet* 344 (8914): 39-40.

Frei, B., and Higdon, J. V. 2003. "Antioxidant activity of tea polyphenols in vivo: evidence from animal studies." *The Journal of Nutrition* 133 (10): 3275S-3284S.

Friedman, M., and Levin, C. E. 1989. "Composition of jimson weed (Datura stramonium) seeds." *Journal of Agricultural and Food Chemistry* 37 (4): 998-1005.

Fukudo, S., Kaneko, H., Akiho, H., Inamori, M., Endo, Y., Okumura, T., Kanazawa, M., Kamiya, T., Sato, K., and Chiba, T. 2015. "Evidence-based clinical practice guidelines for irritable bowel syndrome." *Journal of Gastroenterology* 50 (1): 11-30.

Fukui, H. 2015. "Gut-liver axis in liver cirrhosis: How to manage leaky gut and endotoxemia." *World Journal of Hepatology* 7 (3): 425.

Fukui, H. 2017. "Gut Microbiome-based Therapeutics in Liver Cirrhosis: Basic Consideration for the Next Step." *Journal of Clinical and Translational Hepatology* 5 (3): 249-260.

Furrie, E., Macfarlane, S., Kennedy, A., Cummings, J., Walsh, S., O'neil, D., and Macfarlane, G. 2005. "Synbiotic therapy (Bifidobacterium longum/Synergy 1) initiates resolution of inflammation in patients with active ulcerative colitis: a randomised controlled pilot trial." *Gut* 54 (2): 242-249.

Garcovich, M., Zocco, M. A., Roccarina, D., Ponziani, F. R., and Gasbarrini, A. 2012. "Prevention and treatment of hepatic encephalopathy: focusing on gut microbiota." *World Journal of Gastroenterology: WJG* 18 (46): 6693.

Gibson, G., and Wang, X. 1994. "Regulatory effects of bifidobacteria on the growth of other colonic bacteria." *Journal of Applied Microbiology* 77 (4): 412-420.

Gibson, G. R., Beatty, E. R., Wang, X., and Cummings, J. H. 1995. "Selective stimulation of bifidobacteria in the human colon by oligofructose and inulin." *Gastroenterology* 108 (4): 975-982.

Gibson, G. R., Probert, H. M., Van Loo, J., Rastall, R. A., and Roberfroid, M. B. 2004. "Dietary modulation of the human colonic microbiota: updating the concept of prebiotics." *Nutrition Research Reviews* 17 (2): 259-275.

Gibson, G. R., and Roberfroid, M. B. 1995. "Dietary modulation of the human colonic microbiota: introducing the concept of prebiotics." *The Journal of Nutrition* 125 (6): 1401-1412.

Gómez-Zorita, S., Fernández-Quintela, A., Macarulla, M., Aguirre, L., Hijona, E., Bujanda, L., Milagro, F., Martínez, J., and Portillo, M. 2012. "Resveratrol attenuates steatosis in obese Zucker rats by decreasing fatty acid availability and reducing oxidative stress." *British Journal of Nutrition* 107 (2): 202-210.

Gorbach, S. L. 2000. "Probiotics and gastrointestinal health." *The American Journal of Gastroenterology* 95 (1): S2-S4.

Grant, K. L., and Schneider, C. D. 2000. "Alternative therapies." *American Journal of Health-System Pharmacy* 57 (12): 1121-1122.

Greenstein, R. J. 2003. "Is Crohn's disease caused by a mycobacterium? Comparisons with leprosy, tuberculosis, and Johne's disease." *The Lancet Infectious Diseases* 3 (8): 507-514.

Gridley, T., and Kajimura, S. 2014. "Lightening up a notch: Notch regulation of energy metabolism." *Nature Medicine* 20 (8): 811-812.

Guan, F., Liu, A. B., Li, G., Yang, Z., Sun, Y., Yang, C. S., and Ju, J. 2012. "Deleterious effects of high concentrations of (-)-epigallocatechin-3-gallate and atorvastatin in mice with colon inflammation." *Nutrition and Cancer* 64 (6): 847-855.

Guandalini, S. 2011. "Probiotics for prevention and treatment of diarrhea." *Journal of Clinical Gastroenterology* 45: S149-S153.

Gudbrandsen, O. A., Wergedahl, H., Mork, S., Liaset, B., Espe, M., and Berge, R. K. 2006. "Dietary soya protein concentrate enriched with isoflavones reduced fatty liver, increased hepatic fatty acid oxidation and decreased the hepatic mRNA level of VLDL receptor in obese Zucker rats." *British Journal of Nutrition* 96 (2): 249-57.

Guglielmetti, S., Mora, D., Gschwender, M., and Popp, K. 2011. "Randomised clinical trial: Bifidobacterium bifidum MIMBb75 significantly alleviates irritable bowel syndrome and improves quality of life—a double-blind, placebo-controlled study." *Alimentary Pharmacology & Therapeutics* 33 (10): 1123-1132.

Gukovsky, I., Reyes, C. N., Vaquero, E. C., Gukovskaya, A. S., and Pandol, S. J. 2003. "Curcumin ameliorates ethanol and nonethanol experimental pancreatitis." *American Journal of Physiology-Gastrointestinal and Liver Physiology* 284 (1): G85-G95.

Gul, K., Singh, A., and Jabeen, R. 2016. "Nutraceuticals and functional foods: The foods for the future world." *Critical Reviews in Food Science and Nutrition* 56 (16): 2617-2627.

Gunnarsdottir, S. A., Sadik, R., Shev, S., Simrén, M., Sjövall, H., Stotzer, P.-O., Abrahamsson, H., Olsson, R., and Björnsson, E. S. 2003. "Small intestinal motility disturbances and bacterial overgrowth in patients with liver cirrhosis and portal hypertension." *The American Journal of Gastroenterology* 98 (6): 1362-1370.

Guo, H., Liu, G., Zhong, R., Wang, Y., Wang, D., and Xia, M. 2012. "Cyanidin-3-O-β-glucoside regulates fatty acid metabolism via an AMP-activated protein kinase-dependent signaling pathway in human HepG2 cells." *Lipids in Health and Disease* 11 (1): 10.

Gupta, A., and Rath, P. C. 2014. "Curcumin, a natural antioxidant, acts as a noncompetitive inhibitor of human RNase L in presence of its cofactor 2-5A in vitro." *BioMed Research International* 2014.

Hämäläinen, M., Nieminen, R., Asmawi, M. Z., Vuorela, P., Vapaatalo, H., and Moilanen, E. 2011. "Effects of flavonoids on prostaglandin E2 production and on COX-2 and mPGES-1 expressions in activated macrophages." *Planta Medica* 77 (13): 1504-1511.

Han, J.-J., Hao, J., Kim, C.-H., Hong, J.-S., Ahn, H.-Y., and Lee, Y.-S. 2009. "Quercetin prevents cardiac hypertrophy induced by pressure overload in rats." *Journal of Veterinary Medical Science* 71 (6): 737-743.

Hanai, H., Iida, T., Takeuchi, K., Watanabe, F., Maruyama, Y., Andoh, A., Tsujikawa, T., Fujiyama, Y., Mitsuyama, K., and Sata, M. 2006. "Curcumin maintenance therapy for ulcerative colitis: randomized, multicenter, double-blind, placebo-controlled trial." *Clinical Gastroenterology and Hepatology* 4 (12): 1502-1506.

Hansen, R., Russell, R. K., Reiff, C., Louis, P., McIntosh, F., Berry, S. H., Mukhopadhya, I., Bisset, W. M., Barclay, A. R., and Bishop, J. 2012. "Microbiota of de-novo pediatric IBD: increased Faecalibacterium prausnitzii and reduced bacterial diversity in Crohn's but not in ulcerative colitis." *The American Journal of Gastroenterology* 107 (12): 1913.

Harris, L. A., and Baffy, N. 2017. "Modulation of the gut microbiota: a focus on treatments for irritable bowel syndrome." *Postgraduate Medicine* 129 (8): 872-888.

Hasan, S., Zingg, J.-M., Kwan, P., Noble, T., Smith, D., and Meydani, M. 2014. "Curcumin modulation of high fat diet-induced atherosclerosis and steatohepatosis in LDL receptor deficient mice." *Atherosclerosis* 232 (1): 40-51.

Heber, D., and Lu, Q.-Y. 2002. "Overview of mechanisms of action of lycopene." *Experimental Biology and Medicine* 227 (10): 920-923.

Hecker, M., Preiß, C., Klemm, P., and Busse, R. 1996. "Inhibition by antioxidants of nitric oxide synthase expression in murine macrophages: role of nuclear factor kB and interferon regulatory factor 1." *British Journal of Pharmacology* 118 (8): 2178-2184.

Heeboll, S., El-Houri, R. B., Hellberg, Y. E., Haldrup, D., Pedersen, S. B., Jessen, N., Christensen, L. P., and Gronbaek, H. 2016. "Effect of resveratrol on experimental non-alcoholic fatty liver disease depends on severity of pathology and timing of treatment." *Journal of Gastroenterology and Hepatology* 31 (3): 668-75.

Heeboll, S., Kreuzfeldt, M., Hamilton-Dutoit, S., Kjær Poulsen, M., Stødkilde-Jørgensen, H., Møller, H. J., Jessen, N., Thorsen, K., Kristina Hellberg, Y., and Bønløkke Pedersen, S. 2016. "Placebo-controlled, randomised clinical trial: high-dose resveratrol treatment for non-alcoholic fatty liver disease." *Scandinavian Journal of Gastroenterology* 51 (4): 456-464.

Hempel, S., Newberry, S. J., Maher, A. R., Wang, Z., Miles, J. N., Shanman, R., Johnsen, B., and Shekelle, P. G. 2012. "Probiotics for the prevention and treatment of antibiotic-associated diarrhea: a systematic review and meta-analysis." *Jama* 307 (18): 1959-1969.

Hernández-Ortega, L. D., Alcántar-Díaz, B. E., Ruiz-Corro, L. A., Sandoval-Rodriguez, A., Bueno-Topete, M., Armendariz-Borunda, J., and Salazar-Montes, A. M. 2012. "Quercetin improves hepatic fibrosis reducing hepatic stellate cells and regulating pro-fibrogenic/anti-fibrogenic molecules balance." *Journal of Gastroenterology and Hepatology* 27 (12): 1865-1872.

Hirsch, N., Konstantinov, A., Anavi, S., Aronis, A., Hagay, Z., Madar, Z., and Tirosh, O. 2016. "Prolonged feeding with green tea polyphenols exacerbates cholesterol-induced fatty liver disease in mice." *Molecular Nutrition and Food Research* 60 (12): 2542-2553.

Hoek-van den Hil, E. F., Keijer, J., Bunschoten, A., Vervoort, J. J., Stankova, B., Bekkenkamp, M., Herreman, L., Venema, D., Hollman, P. C., and Tvrzicka, E. 2013. "Quercetin induces hepatic lipid omega-oxidation and lowers serum lipid levels in mice." *PloS one* 8 (1): e51588.

Hofacre, C., Mathis, G., and Quiroz, M. 2005. "Natural alternatives to prevent necrotic enteritis." *Int. Poult. Prod* 13: 7-9.

Holt, P. R., Katz, S., and Kirshoff, R. 2005. "Curcumin therapy in inflammatory bowel disease: a pilot study." *Digestive Diseases and Sciences* 50 (11): 2191-2193.

Hosein, F. M., Abbasabadi, Z., Reza, S.-A. M., Abdollahi, M., and Rahimi, R. 2014. "A comprehensive review of plants and their active constituents with wound healing activity in traditional Iranian medicine." *Wounds: A Compendium of Clinical Research and Practice* 26 (7): 197-206.

Hou, J. K., Abraham, B., and El-Serag, H. 2011. "Dietary intake and risk of developing inflammatory bowel disease: a systematic review of the literature." *The American Journal of Gastroenterology* 106 (4): 563.

Houdeau, E., Moriez, R., Leveque, M., Salvador–Cartier, C., Waget, A., Leng, L., Bueno, L., Bucala, R., and Fioramonti, J. 2007. "Sex steroid regulation of macrophage migration inhibitory factor in normal and inflamed colon in the female rat." *Gastroenterology* 132 (3): 982-993.

Houghton, L. A., Lea, R., Jackson, N., and Whorwell, P. 2002. "The menstrual cycle affects rectal sensitivity in patients with irritable bowel syndrome but not healthy volunteers." *Gut* 50 (4): 471-474.

Houglum, K., Filip, M., Witztum, J. L., and Chojkier, M. 1990. "Malondialdehyde and 4-hydroxynonenal protein adducts in plasma and liver of rats with iron overload." *The Journal of Clinical Investigation* 86 (6): 1991-1998.

Hubert, J., Berger, M., Nepveu, F., Paul, F., and Daydé, J. 2008. "Effects of fermentation on the phytochemical composition and antioxidant properties of soy germ." *Food Chemistry* 109 (4): 709-721.

Igarashi, M., Nagano, J., Tsuda, A., Suzuki, T., Koike, J., Uchida, T., Matsushima, M., Mine, T., and Koga, Y. 2014. "Correlation between the Serum Pepsinogen I Level and the Symptom Degree in Proton Pump Inhibitor-Users Administered with a Probiotic." *Pharmaceuticals (Basel)* 7 (7): 754-64.

Imai, K., and Nakachi, K. 1995. "Cross sectional study of effects of drinking green tea on cardiovascular and liver diseases." *Bmj* 310 (6981): 693-696.

Ishikawa, H., Akedo, I., Umesaki, Y., Tanaka, R., Imaoka, A., and Otani, T. 2003. "Randomized controlled trial of the effect of bifidobacteria-fermented milk on ulcerative colitis." *Journal of the American College of Nutrition* 22 (1): 56-63.

Isolauri, E., Salminen, S., and Ouwehand, A. C. 2004. "Probiotics." *Best Practice & Research Clinical Gastroenterology* 18 (2): 299-313.

Isolauri, E., Sütas, Y., Kankaanpää, P., Arvilommi, H., and Salminen, S. 2001. "Probiotics: effects on immunity." *The American Journal of Clinical Nutrition* 73 (2): 444s-450s.

Jeon, B. T., Jeong, E. A., Shin, H. J., Lee, Y., Lee, D. H., Kim, H. J., Kang, S. S., Cho, G. J., Choi, W. S., and Roh, G. S. 2012. "Resveratrol attenuates obesity-associated peripheral and central inflammation and improves memory deficit in mice fed a high-fat diet." *Diabetes* 61 (6): 1444-1454.

Jeon, H. J., Seo, M. J., Choi, H. S., Lee, O. H., and Lee, B. Y. 2014. "Gelidium elegans, an Edible Red Seaweed, and Hesperidin Inhibit Lipid Accumulation and Production of Reactive Oxygen Species and Reactive Nitrogen Species in 3T3-L1 and RAW264.7 Cells." *Phytotherapy Research* 28 (11): 1701-1709.

Joe, B., Vijaykumar, M., and Lokesh, B. 2004. "Biological properties of curcumin-cellular and molecular mechanisms of action." *Critical Reviews in Food Science and Nutrition* 44 (2): 97-111.

Joerg, G., Thomas, B., Christof, J., and Mukesh, M. 2000. "In; PDR for Herbal Medicines." *Medical Economics Company, Montvale, New Jersey* 319.

Johnston, B. C., Supina, A. L., Ospina, M., and Vohra, S. 2007. "Probiotics for the prevention of pediatric antibiotic-associated diarrhea." *Cochrane Database of Systematic Reviews* 18 (2): 4827.

Jones, J. R., Lineback, D. M., and Levine, M. J. 2006. "Dietary reference intakes: implications for fiber labeling and consumption: a summary of the International Life Sciences Institute North America Fiber Workshop, June 1–2, 2004, Washington, DC." *Nutrition Reviews* 64 (1): 31-38.

Jump, D. B. 2008. "N-3 polyunsaturated fatty acid regulation of hepatic gene transcription." *Current Opinion in Lipidology* 19 (3): 242-247.

Jung, C. H., Cho, I., Ahn, J., Jeon, T. I., and Ha, T. Y. 2013. "Quercetin reduces high-fat diet-induced fat accumulation in the liver by regulating lipid metabolism genes." *Phytotherapy Research* 27 (1): 139-143.

Kahraman, A., Erkasap, N., Köken, T., Serteser, M., Aktepe, F., and Erkasap, S. 2003. "The antioxidative and antihistaminic properties of quercetin in ethanol-induced gastric lesions." *Toxicology* 183 (1): 133-142.

Kaithwas, G., Mukerjee, A., Kumar, P., and Majumdar, D. K. 2011. "Linum usitatissimum (linseed/flaxseed) fixed oil: antimicrobial activity and efficacy in bovine mastitis." *Inflammopharmacology* 19 (1): 45-52.

Kamimura, S., Gaal, K., Britton, R. S., Bacon, B. R., Triadafilopoulos, G., and Tsukamoto, H. 1992. "Increased 4-hydroxynonenal levels in experimental alcoholic liver disease: Association of lipid peroxidation with liver fibrogenesis." *Hepatology* 16 (2): 448-453.

Kang, H. C., Nan, J. X., Park, P. H., Kim, J. Y., Lee, S. H., Woo, S. W., Zhao, Y. Z., Park, E. J., and Sohn, D. H. 2002. "Curcumin inhibits collagen synthesis and hepatic stellate cell activation in-vivo and in-vitro." *Journal of Pharmacy and Pharmacology* 54 (1): 119-126.

Kang, J. W., and Lee, S. M. 2014. "Protective Effects of Chlorogenic Acid against Experimental Reflux Esophagitis in Rats." *Biomolecules and Therapeutics (Seoul)* 22 (5): 420-5.

Karell, K., Louka, A. S., Moodie, S. J., Ascher, H., Clot, F., Greco, L., Ciclitira, P. J., Sollid, L. M., Partanen, J., and Disease, E. G. C. o. C. 2003. "HLA types in celiac disease patients not carrying the DQA1* 05-DQB1* 02 (DQ2) heterodimer: results from the European Genetics Cluster on Celiac Disease." *Human Immunology* 64 (4): 469-477.

Kassinen, A., Krogius-Kurikka, L., Mäkivuokko, H., Rinttilä, T., Paulin, L., Corander, J., Malinen, E., Apajalahti, J., and Palva, A. 2007. "The fecal microbiota of irritable bowel syndrome patients differs significantly from that of healthy subjects." *Gastroenterology* 133 (1): 24-33.

Kaulmann, A., and Bohn, T. 2014. "Carotenoids, inflammation, and oxidative stress—implications of cellular signaling pathways and relation to chronic disease prevention." *Nutrition Research* 34 (11): 907-929.

Kaulmann, A., and Bohn, T. 2016. "Bioactivity of polyphenols: Preventive and adjuvant strategies toward reducing inflammatory bowel diseases—promises, perspectives, and pitfalls." *Oxidative Medicine and Cellular Longevity* 2016: 9346470.

Kawahara, H., Kubota, A., Hasegawa, T., Okuyama, H., Ueno, T., Ida, S., and Fukuzawa, M. 2007. "Effects of rikkunshito on the clinical symptoms and esophageal acid exposure in children with symptomatic gastroesophageal reflux." *Pediatric Surgery International* 23 (10): 1001-1005.

Keely, S. J., Scharl, M. M., Bertelsen, L. S., Hagey, L. R., Barrett, K. E., and Hofmann, A. F. 2007. "Bile acid-induced secretion in polarized monolayers of T84 colonic epithelial cells: Structure-activity relationships." *American Journal of Physiology-Gastrointestinal and Liver Physiology* 292 (1): G290-G297.

Kelley, D. S., Vemuri, M., Adkins, Y., Gill, S. H. S., Fedor, D., and Mackey, B. E. 2008. "Flaxseed oil prevents trans-10, cis-12-conjugated linoleic acid-induced insulin resistance in mice." *British Journal of Nutrition* 101 (5): 701-708.

Kelly, C. P., and LaMont, J. T. 2008. "Clostridium difficile—more difficult than ever." *New England Journal of Medicine* 359 (18): 1932-1940.

Khalif, I., Quigley, E., Konovitch, E., and Maximova, I. 2005. "Alterations in the colonic flora and intestinal permeability and evidence of immune activation in chronic constipation." *Digestive and Liver Disease* 37 (11): 838-849.

Khonche, A., Biglarian, O., Panahi, Y., Valizadegan, G., Soflaei, S. S., Ghamarchehreh, M. E., Majeed, M., and Sahebkar, A. 2016. "Adjunctive Therapy with Curcumin for Peptic Ulcer: a Randomized Controlled Trial." *Drug Research (Stuttg)* 66 (8): 444-8.

Kim, D.-C., Kim, S.-H., Choi, B.-H., Baek, N.-I., Kim, D., Kim, M.-J., and Kim, K.-T. 2005. "Curcuma longa extract protects against gastric ulcers by blocking H2 histamine receptors." *Biological and Pharmaceutical Bulletin* 28 (12): 2220-2224.

Kim, G. W., Lin, J. E., Blomain, E. S., and Waldman, S. A. 2014. "Antiobesity pharmacotherapy: new drugs and emerging targets." *Clinical Pharmacology & Therapeutics* 95 (1): 53-66.

Kim, H., Vazquez Roque, M., Camilleri, M., Stephens, D., Burton, D., Baxter, K., Thomforde, G., and Zinsmeister, A. R. 2005. "A randomized controlled trial of a probiotic combination VSL# 3 and placebo in irritable bowel syndrome with bloating." *Neurogastroenterology & Motility* 17 (5): 687-696.

Kim, M.-H., Kang, K.-S., and Lee, Y.-S. 2010. "The inhibitory effect of genistein on hepatic steatosis is linked to visceral adipocyte metabolism in mice with diet-induced non-alcoholic fatty liver disease." *British Journal of Nutrition* 104 (9): 1333-1342.

Kim, M., Park, J., Jung, J., Byun, K., Kang, K., and Lee, Y. 2011. "Daidzein supplementation prevents non-alcoholic fatty liver disease through alternation of hepatic gene expression profiles and adipocyte metabolism." *International Journal of Obesity* 35 (8): 1019-1030.

Kim, S., Shin, H.-J., Kim, S. Y., Kim, J. H., Lee, Y. S., Kim, D.-H., and Lee, M.-O. 2004. "Genistein enhances expression of genes involved in fatty acid catabolism through activation of PPARα." *Molecular and Cellular Endocrinology* 220 (1): 51-58.

Kincaid, B., and Bossy-Wetzel, E. 2013. "Forever young: SIRT3 a shield against mitochondrial meltdown, aging, and neurodegeneration." *Frontiers in Aging Neuroscience* 5: 48.

Knipping, K., Garssen, J., and van't Land, B. 2012. "An evaluation of the inhibitory effects against rotavirus infection of edible plant extracts." *Virology Journal* 9 (1): 1.

Kobori, M., Masumoto, S., Akimoto, Y., and Oike, H. 2011. "Chronic dietary intake of quercetin alleviates hepatic fat accumulation associated with consumption of a Western-style diet in C57/BL6J mice." *Molecular Nutrition & Food Research* 55 (4): 530-540.

Koebnick, C., Wagner, I., Leitzmann, P., Stern, U., and Zunft, H. 2003. "Probiotic beverage containing Lactobacillus casei Shirota improves gastrointestinal symptoms in patients with chronic constipation." *Canadian Journal of Gastroenterology and Hepatology* 17 (11): 655-659.

Kubo, A., Block, G., Quesenberry, C. P., Jr., Buffler, P., and Corley, D. A. 2009. "Effects of dietary fiber, fats, and meat intakes on the risk of Barrett's esophagus." *Nutrition and Cancer* 61 (5): 607-16.

Kuehbacher, T., Ott, S. J., Helwig, U., Mimura, T., Rizzello, F., Kleessen, B., Gionchetti, P., Blaut, M., Campieri, M., and Fölsch, U. R. 2006. "Bacterial and fungal microbiota in relation to probiotic therapy (VSL# 3) in pouchitis." *Gut* 55 (6): 833-841.

Kuiper, G. G., Lemmen, J. G., Carlsson, B., Corton, J. C., Safe, S. H., Van Der Saag, P. T., Van Der Burg, B., and Gustafsson, J.-A. k. 1998. "Interaction of estrogenic chemicals and phytoestrogens with estrogen receptor β." *Endocrinology* 139 (10): 4252-4263.

Lacy, B. E., Chey, W. D., and Lembo, A. J. 2015. "New and emerging treatment options for irritable bowel syndrome." *Gastroenterology & Hepatology* 11 (4 Suppl 2): 1.

Lagouge, M., Argmann, C., Gerhart-Hines, Z., Meziane, H., Lerin, C., Daussin, F., Messadeq, N., Milne, J., Lambert, P., and Elliott, P. 2006. "Resveratrol improves mitochondrial function and protects against metabolic disease by activating SIRT1 and PGC-1α." *Cell* 127 (6): 1109-1122.

Lammers, K. M., Brigidi, P., Vitali, B., Gionchetti, P., Rizzello, F., Caramelli, E., Matteuzzi, D., and Campieri, M. 2003. "Immunomodulatory effects of probiotic bacteria DNA: IL-1 and IL-10 response in human peripheral blood mononuclear cells." *FEMS Immunology & Medical Microbiology* 38 (2): 165-172.

Lang, L., Qi, L., Zhang, J., Shi, N., Zhang, M., Wu, T., and Chen, J. 2001. "Effect of genistein and quercetin on proliferation, collagen synthesis, and type I procollagen mRNA levels of rat hepatic stellate cells." *Acta Pharmacologica Sinica* 22 (9): 793-796.

Langen, L. V., Mirjam, A., and Dieleman, L. A. 2009. "Prebiotics in chronic intestinal inflammation." *Inflammatory Bowel Diseases* 15 (3): 454-462.

Leclercq, I. A., Farrell, G. C., Sempoux, C., dela Peña, A., and Horsmans, Y. 2004. "Curcumin inhibits NF-κB activation and reduces the severity of experimental steatohepatitis in mice." *Journal of Hepatology* 41 (6): 926-934.

Lee, K. S., Buck, M., Houglum, K., and Chojkier, M. 1995. "Activation of hepatic stellate cells by TGF alpha and collagen type I is mediated by oxidative stress through c-myb expression." *The Journal of Clinical Investigation* 96 (5): 2461-2468.

Lee, M.-S., Kim, C.-T., and Kim, Y. 2009. "Green tea (–)-epigallocatechin-3-gallate reduces body weight with regulation of multiple genes expression in adipose tissue of diet-induced obese mice." *Annals of Nutrition and Metabolism* 54 (2): 151-157.

Levy, J. 2000. "The effects of antibiotic use on gastrointestinal function." *The American Journal of Gastroenterology* 95 (1): S8-S10.

Li, L., Hai, J., Li, Z., Zhang, Y., Peng, H., Li, K., and Weng, X. 2014. "Resveratrol modulates autophagy and NF-κB activity in a murine model for treating non-alcoholic fatty liver disease." *Food and Chemical Toxicology* 63: 166-173.

Li, Y., Xu, S., Mihaylova, M. M., Zheng, B., Hou, X., Jiang, B., Park, O., Luo, Z., Lefai, E., and Shyy, J. Y.-J. 2011. "AMPK phosphorylates and inhibits SREBP activity to attenuate hepatic steatosis and atherosclerosis in diet-induced insulin-resistant mice." *Cell Metabolism* 13 (4): 376-388.

Liu, Q., Duan, Z. P., Ha, D. K., Bengmark, S., Kurtovic, J., and Riordan, S. M. 2004. "Synbiotic modulation of gut flora: effect on minimal hepatic encephalopathy in patients with cirrhosis." *Hepatology* 39 (5): 1441-1449.

Loguercio, C., Federico, A., Tuccillo, C., Terracciano, F., D'auria, M. V., De Simone, C., and Blanco, C. D. V. 2005. "Beneficial effects of a probiotic VSL# 3 on parameters of liver dysfunction in chronic liver diseases." *Journal of Clinical Gastroenterology* 39 (6): 540-543.

Longstreth, G. F. 2005. "Definition and classification of irritable bowel syndrome: current consensus and controversies." *Gastroenterology Clinics of North America* 34 (2): 173-187.

Lu, M., Yu, S., Deng, J., Yan, Q., Yang, C., Xia, G., and Zhou, X. 2016. "Efficacy of Probiotic Supplementation Therapy for Helicobacter pylori Eradication: A Meta-Analysis of Randomized Controlled Trials." *PLoS One* 11 (10): e0163743.

Lunia, M. K., Sharma, B. C., Sharma, P., Sachdeva, S., and Srivastava, S. 2014. "Probiotics prevent hepatic encephalopathy in patients with cirrhosis: a randomized controlled trial." *Clinical Gastroenterology and Hepatology* 12 (6): 1003-1008. e1.

Ma, F., Zhou, C., Wang, J., Liu, T., and Liu, J. 2015. "Probiotics in the treatment of peptic ulcer infected by helicobacter pylory and its safety." *Pakistan Journal of Pharmaceutical Sciences* 28 (3 Suppl): 1087-90.

Ma, Y. Y., Li, L., Yu, C. H., Shen, Z., Chen, L. H., and Li, Y. M. 2013. "Effects of probiotics on nonalcoholic fatty liver disease: a meta-analysis." *World Journal of Gastroenterology* 19 (40): 6911-8.

Mack, D. R., Michail, S., Wei, S., McDougall, L., and Hollingsworth, M. A. 1999. "Probiotics inhibit enteropathogenic E. coliadherence in vitro by inducing intestinal mucin gene expression." *American Journal of Physiology-Gastrointestinal and Liver Physiology* 276 (4): G941-G950.

Mackenzie, P. I., Bock, K. W., Burchell, B., Guillemette, C., Ikushiro, S.-i., Iyanagi, T., Miners, J. O., Owens, I. S., and Nebert, D. W. 2005. "Nomenclature update for the mammalian UDP glycosyltransferase (UGT) gene superfamily." *Pharmacogenetics and Genomics* 15 (10): 677-685.

Mahady, G. B., Pendland, S. L., and Chadwick, L. R. 2003. "Resveratrol and red wine extracts inhibit the growth of CagA+ strains of Helicobacter pylori in vitro." *The American Journal of Gastroenterology* 98 (6): 1440.

Mahattanadul, S., Ridtitid, W., Nima, S., Phdoongsombut, N., Ratanasuwon, P., and Kasiwong, S. 2011. "Effects of Morinda citrifolia aqueous fruit extract and its biomarker scopoletin on reflux esophagitis and gastric ulcer in rats." *Journal of Ethnopharmacology* 134 (2): 243-250.

Malaguarnera, M., Gargante, M. P., Malaguarnera, G., Salmeri, M., Mastrojeni, S., Rampello, L., Pennisi, G., Volti, G. L., and Galvano, F. 2010. "Bifidobacterium combined with fructo-oligosaccharide versus lactulose in the treatment of patients with hepatic encephalopathy." *European Journal of Gastroenterology & Hepatology* 22 (2): 199-206.

Malasanos, T. H., and Stacpoole, P. W. 1991. "Biological effects of ω-3 fatty acids in diabetes mellitus." *Diabetes Care* 14 (12): 1160-1179.

Manach, C., Williamson, G., Morand, C., Scalbert, A., and Rémésy, C. 2005. "Bioavailability and bioefficacy of polyphenols in humans. I. Review of 97 bioavailability studies." *The American Journal of Clinical Nutrition* 81 (1): 230S-242S.

Manco, M., Putignani, L., and Bottazzo, G. F. 2010. "Gut microbiota, lipopolysaccharides, and innate immunity in the pathogenesis of obesity and cardiovascular risk." *Endocrine Reviews* 31 (6): 817-844.

Mariita, R. M., Orodho, J. A., Okemo, P. O., Kirimuhuzya, C., Otieno, J. N., and Magadula, J. J. 2011. "Methanolic extracts of Aloe secundiflora Engl. inhibits in vitro growth of tuberculosis and diarrhea-causing bacteria." *Pharmacognosy Research* 3 (2): 95.

Marotta, R. B., and Floch, M. H. 1991. "Diet and nutrition in ulcer disease." *Medical Clinics of North America* 75 (4): 967-79.

Marra, F., Gastaldelli, A., Baroni, G. S., Tell, G., and Tiribelli, C. 2008. "Molecular basis and mechanisms of progression of non-alcoholic steatohepatitis." *Trends in Molecular Medicine* 14 (2): 72-81.

Martín-Pozuelo, G., Navarro-González, I., González-Barrio, R., Santaella, M., García-Alonso, J., Hidalgo, N., Gómez-Gallego, C., Ros, G., and Periago, M. J. 2015. "The effect of tomato juice supplementation on biomarkers and gene expression related to lipid metabolism in rats with induced hepatic steatosis." *European Journal of Nutrition* 54 (6): 933-944.

Martin, D. A., and Bolling, B. W. 2015. "A review of the efficacy of dietary polyphenols in experimental models of inflammatory bowel diseases." *Food & Function* 6 (6): 1773-1786.

Martinello, F., Roman, C. F., and Souza, P. A. 2017. "Effects of probiotic intake on intestinal bifidobacteria of celiac patients." *Arquivos de Gastroentero-logia* 54 (2): 85-90.

Martínez-Flórez, S., González-Gallego, J., Culebras, J. M., and Tuñón, M. J. 2002. "Los flavonoides: propiedades y acciones antioxidantes." ["Flavonoids: antioxidant properties and actions"]. *Nutricion Hospitalaria* 17 (6): 271-278.

Martínez-Flórez, S., Gutiérrez-Fernández, B., Sánchez-Campos, S., González-Gallego, J., and Tuñón, M. J. 2005. "Quercetin attenuates nuclear factor-κB activation and nitric oxide production in interleukin-1β–activated rat hepatocytes." *The Journal of Nutrition* 135 (6): 1359-1365.

Marx, N., Duez, H., Fruchart, J.-C., and Staels, B. 2004. "Peroxisome proliferator-activated receptors and atherogenesis." *Circulation Research* 94 (9): 1168-1178.

Matsuda, H., Morikawa, T., Ando, S., Toguchida, I., and Yoshikawa, M. 2003. "Structural requirements of flavonoids for nitric oxide production inhibitory activity and mechanism of action." *Bioorganic & Medicinal Chemistry* 11 (9): 1995-2000.

Matysiak-Budnik, T., Candalh, C., Dugave, C., Namane, A., Cellier, C., Cerf-Bensussan, N., and Heyman, M. 2003. "Alterations of the intestinal transport and processing of gliadin peptides in celiac disease." *Gastroenterology* 125 (3): 696-707.

McCartney, A. L., Wenzhi, W., and Tannock, G. W. 1996. "Molecular analysis of the composition of the bifidobacterial and lactobacillus microflora of humans." *Applied and Environmental Microbiology* 62 (12): 4608-4613.

McFarland, L. V. 2006. "Meta-analysis of probiotics for the prevention of antibiotic associated diarrhea and the treatment of Clostridium difficile disease." *The American Journal of Gastroenterology* 101 (4): 812-822.

McGee, R. G., Bakens, A., Wiley, K., Riordan, S. M., and Webster, A. C. 2011. "Probiotics for patients with hepatic encephalopathy." *Cochrane Database of Systematic Reviews* 11: CD008716.

McKenzie, Y., Bowyer, R., Leach, H., Gulia, P., Horobin, J., O'sullivan, N., Pettitt, C., Reeves, L., Seamark, L., and Williams, M. 2016. "British Dietetic Association systematic review and evidence-based practice guidelines for the dietary management of irritable bowel syndrome in adults (2016 update)." *Journal of Human Nutrition and Dietetics* 29 (5): 549-575.

McRorie, J., Kesler, J., Bishop, L., Filloon, T., Allgood, G., Sutton, M., Hunt, T., Laurent, A., and Rudolph, C. 2000. "Effects of wheat bran and Olestra on objective measures of stool and subjective reports of GI symptoms." *The American Journal of Gastroenterology* 95 (5): 1244.

McRorie Jr, J. W. 2015. "Evidence-based approach to fiber supplements and clinically meaningful health benefits, part 1: What to look for and how to recommend an effective fiber therapy." *Nutrition Today* 50 (2): 82.

Mennigen, R., Nolte, K., Rijcken, E., Utech, M., Loeffler, B., Senninger, N., and Bruewer, M. 2009. "Probiotic mixture VSL# 3 protects the epithelial barrier by maintaining tight junction protein expression and preventing apoptosis in a murine model of colitis." *American Journal of Physiology-Gastrointestinal and Liver Physiology* 296 (5): G1140-G1149.

Miele, E., Pascarella, F., Giannetti, E., Quaglietta, L., Baldassano, R. N., and Staiano, A. 2009. "Effect of a probiotic preparation (VSL# 3) on induction and maintenance of remission in children with ulcerative colitis." *The American Journal of Gastroenterology* 104 (2): 437.

Minemura, M., and Shimizu, Y. 2015. "Gut microbiota and liver diseases." *World Journal of Gastroenterology* 21 (6): 1691-1702.

Mittal, V. V., Sharma, B. C., Sharma, P., and Sarin, S. K. 2011. "A randomized controlled trial comparing lactulose, probiotics, and L-ornithine L-aspartate in treatment of minimal hepatic encephalopathy." *European Journal of Gastroenterology & Hepatology* 23 (8): 725-732.

Miyoshi, M., Sakaki, H., Usami, M., Iizuka, N., Shuno, K., Aoyama, M., and Usami, Y. 2011. "Oral administration of tributyrin increases concentration of butyrate in the portal vein and prevents lipopolysaccharide-induced liver injury in rats." *Clinical Nutrition* 30 (2): 252-258.

Moayyedi, P., Quigley, E. M., Lacy, B. E., Lembo, A. J., Saito, Y. A., Schiller, L. R., Soffer, E. E., Spiegel, B. M., and Ford, A. C. 2014. "The effect of fiber supplementation on irritable bowel syndrome: a systematic review and meta-analysis." *The American Journal of Gastroenterology* 109 (9): 1367.

Mohammadi, A., Sahebkar, A., Iranshahi, M., Amini, M., Khojasteh, R., Ghayour-Mobarhan, M., and Ferns, G. A. 2013. "Effects of supplementation with curcuminoids on dyslipidemia in obese patients: a randomized crossover trial." *Phytotherapy Research* 27 (3): 374-379.

Morimoto, M., Watanabe, T., Yamori, M., Takebe, M., and Wakatsuki, Y. 2009. "Isoflavones regulate innate immunity and inhibit experimental colitis." *Journal of Gastroenterology and Hepatology* 24 (6): 1123-1129.

Morito, K., Hirose, T., Kinjo, J., Hirakawa, T., Okawa, M., Nohara, T., Ogawa, S., Inoue, S., Muramatsu, M., and Masamune, Y. 2001. "Interaction of phytoestrogens with estrogen receptors α and β." *Biological and Pharmaceutical Bulletin* 24 (4): 351-356.

Morris Brown, L., Swanson, C. A., Gridley, G., Swanson, G. M., Schoenberg, J. B., Greenberg, R. S., Silverman, D. T., Pottern, L. M., Hayes, R. B., and Schwartz, A. G. 1995. "Adenocarcinoma of the esophagus: role of obesity and diet." *JNCI: Journal of the National Cancer Institute* 87 (2): 104-109.

Morris, D. H. 2007. *Flax: A Health and Nutrition Primer*. Flax Council of Canada.

Moussa, L., Bézirard, V., Salvador-Cartier, C., Bacquié, V., Houdeau, E., and Théodorou, V. 2013. "A new soy germ fermented ingredient displays estrogenic and protease inhibitor activities able to prevent irritable bowel syndrome-like symptoms in stressed female rats." *Clinical Nutrition* 32 (1): 51-58.

Moussa, L., Bézirard, V., Salvador-Cartier, C., Bacquié, V., Lencina, C., Lévêque, M., Braniste, V., Ménard, S., Théodorou, V., and Houdeau, E. 2012. "A low dose of fermented soy germ alleviates gut barrier injury, hyperalgesia and faecal protease activity in a rat model of inflammatory bowel disease." *PloS one* 7 (11): e49547.

Muccioli, G. G., Naslain, D., Bäckhed, F., Reigstad, C. S., Lambert, D. M., Delzenne, N. M., and Cani, P. D. 2010. "The endocannabinoid system links gut microbiota to adipogenesis." *Molecular Systems Biology* 6 (1): 392.

Mulholland, H. G., Cantwell, M. M., Anderson, L. A., Johnston, B. T., Watson, R. G., Murphy, S. J., Ferguson, H. R., McGuigan, J., Reynolds, J. V., Comber, H., and Murray, L. J. 2009. "Glycemic index, carbohydrate and fiber intakes and risk of reflux esophagitis, Barrett's esophagus, and esophageal adenocarcinoma." *Cancer Causes Control* 20 (3): 279-88.

Münzenmaier, A., Lange, C., Glocker, E., Covacci, A., Moran, A., Bereswill, S., Baeuerle, P. A., Kist, M., and Pahl, H. L. 1997. "A secreted/shed product of Helicobacter pylori activates transcription factor nuclear factor-kappa B." *The Journal of Immunology* 159 (12): 6140-6147.

Na, L. X., Li, Y., Pan, H. Z., Zhou, X. L., Sun, D. J., Meng, M., Li, X. X., and Sun, C. H. 2013. "Curcuminoids exert glucose-lowering effect in type 2 diabetes by decreasing serum free fatty acids: a double-blind, placebo-controlled trial." *Molecular Nutrition & Food Research* 57 (9): 1569-1577.

Nabavi, S., Rafraf, M., Somi, M., Homayouni-Rad, A., and Asghari-Jafarabadi, M. 2014. "Effects of probiotic yogurt consumption on metabolic factors in individuals with nonalcoholic fatty liver disease." *Journal of Dairy Science* 97 (12): 7386-7393.

Nadal, I., Donant, E., Ribes-Koninckx, C., Calabuig, M., and Sanz, Y. 2007. "Imbalance in the composition of the duodenal microbiota of children with coeliac disease." *Journal of Medical Microbiology* 56 (12): 1669-1674.

Nagarajan, N., Morden, A., Bischof, D., King, E. A., Kosztowski, M., Wick, E. C., and Stein, E. M. 2015. "The role of fiber supplementation in the treatment of irritable bowel syndrome: a systematic review and meta-analysis." *European Journal of Gastroenterology & Hepatology* 27 (9): 1002-1010.

Nair, M. R., Chouhan, D., Sen Gupta, S., and Chattopadhyay, S. 2016. "Fermented Foods: Are They Tasty Medicines for Helicobacter pylori Associated Peptic Ulcer and Gastric Cancer?" *Frontiers in Microbiology* 7: 1148.

Nakamoto, K., Takayama, F., Mankura, M., Hidaka, Y., Egashira, T., Ogino, T., Kawasaki, H., and Mori, A. 2009. "Beneficial effects of fermented green tea extract in a rat model of non-alcoholic steatohepatitis." *Journal of Clinical Biochemistry and Nutrition* 44 (3): 239-246.

Nakao, M., Ogura, Y., Satake, S., Ito, I., Iguchi, A., Takagi, K., and Nabeshima, T. 2002. "Usefulness of soluble dietary fiber for the treatment of diarrhea during enteral nutrition in elderly patients." *Nutrition* 18 (1): 35-9.

Nanji, A. A., Jokelainen, K., Tipoe, G. L., Rahemtulla, A., Thomas, P., and Dannenberg, A. J. 2003. "Curcumin prevents alcohol-induced liver disease in rats by inhibiting the expression of NF-κB-dependent genes." *American Journal of Physiology-Gastrointestinal and Liver Physiology* 284 (2): G321-G327.

Nell, S., Suerbaum, S., and Josenhans, C. 2010. "The impact of the microbiota on the pathogenesis of IBD: lessons from mouse infection models." *Nature Reviews Microbiology* 8 (8): 564.

Nguyen, T. L. A., Vieira-Silva, S., Liston, A., and Raes, J. 2015. "How informative is the mouse for human gut microbiota research?" *Disease Models & Mechanisms* 8 (1): 1-16.

Nichols, L. A., Jackson, D. E., Manthey, J. A., Shukla, S. D., and Holland, L. J. 2011. "Citrus flavonoids repress the mRNA for stearoyl-CoA desaturase, a key enzyme in lipid synthesis and obesity control, in rat primary hepatocytes." *Lipids in Health and Disease* 10 (1): 36.

Nicolucci, A. 2015. "Effect of Prebiotic Fiber Intake on Adiposity and Inflammation in Overweight and Obese Children: Assessing the Role of the Gut Microbiota." University of Calgary.

Nistal, E., Caminero, A., Herrán, A. R., Arias, L., Vivas, S., de Morales, J. M. R., Calleja, S., de Miera, L. E. S., Arroyo, P., and Casqueiro, J. 2012. "Differences of small intestinal bacteria populations in adults and children with/without celiac disease: effect of age, gluten diet, and disease." *Inflammatory Bowel Diseases* 18 (4): 649-656.

Njume, C., and Goduka, N. I. 2012. "Treatment of diarrhoea in rural African communities: an overview of measures to maximise the medicinal potentials of indigenous plants." *International Journal of Environmental Research and Public Health* 9 (11): 3911-3933.

Nobaek, S., Johansson, M.-L., Molin, G., Ahrné, S., and Jeppsson, B. 2000. "Alteration of intestinal microflora is associated with reduction in abdominal bloating and pain in patients with irritable bowel syndrome." *The American Journal of Gastroenterology* 95 (5): 1231-1238.

Ohira, H., Fujioka, Y., Katagiri, C., Yano, M., Mamoto, R., Aoyama, M., Usami, M., and Ikeda, M. 2011. "Butyrate enhancement of inteleukin-1β production via activation of oxidative stress pathways in lipopolysaccharide-stimulated THP-1 cells." *Journal of Clinical Biochemistry and Nutrition* 50 (1): 59-66.

Öner-İyidoğan, Y., Koçak, H., Seyidhanoğlu, M., Gürdöl, F., Gülçubuk, A., Yildirim, F., Çevik, A., and Uysal, M. 2013. "Curcumin prevents liver fat accumulation and serum fetuin-A increase in rats fed a high-fat diet." *Journal of Physiology and Biochemistry* 69 (4): 677-686.

Österberg, E., Blomquist, L., Krakau, I., Weinryb, R., Åsberg, M., and Hultcrantz, R. 2000. "A population study on irritable bowel syndrome and mental health." *Scandinavian Journal of Gastroenterology* 35 (3): 264-268.

Ouwehand, A. C., Lagström, H., Suomalainen, T., and Salminen, S. 2002. "Effect of probiotics on constipation, fecal azoreductase activity and fecal mucin content in the elderly." *Annals of nutrition and metabolism* 46 (3-4): 159-162.

Ozturk, F., Gul, M., Ates, B., Ozturk, I. C., Cetin, A., Vardi, N., Otlu, A., and Yilmaz, I. 2009. "Protective effect of apricot (Prunus armeniaca L.) on hepatic steatosis and damage induced by carbon tetrachloride in Wistar rats." *British Journal of Nutrition* 102 (12): 1767-1775.

Palla, A. H., Khan, N. A., Bashir, S., Ur-Rehman, N., Iqbal, J., and Gilani, A. H. 2015. "Pharmacological basis for the medicinal use of Linum usitatissimum (Flaxseed) in infectious and non-infectious diarrhea." *Journal of Ethnopharmacology* 160: 61-8.

Panahi, Y., Alishiri, G. H., Parvin, S., and Sahebkar, A. 2016. "Mitigation of systemic oxidative stress by curcuminoids in osteoarthritis: results of a randomized controlled trial." *Journal of Dietary Supplements* 13 (2): 209-220.

Panahi, Y., Ghanei, M., Hajhashemi, A., and Sahebkar, A. 2016. "Effects of curcuminoids-piperine combination on systemic oxidative stress, clinical symptoms and quality of life in subjects with chronic pulmonary complications due to sulfur mustard: a randomized controlled trial." *Journal of Dietary Supplements* 13 (1): 93-105.

Panahi, Y., Sahebkar, A., Amiri, M., Davoudi, S. M., Beiraghdar, F., Hoseininejad, S. L., and Kolivand, M. 2012. "Improvement of sulphur mustard-induced chronic pruritus, quality of life and antioxidant status by curcumin: results of a randomised, double-blind, placebo-controlled trial." *British Journal of Nutrition* 108 (7): 1272-1279.

Panaro, M. A., Carofiglio, V., Acquafredda, A., Cavallo, P., and Cianciulli, A. 2012. "Anti-inflammatory effects of resveratrol occur via inhibition of lipopolysaccharide-induced NF-κB activation in Caco-2 and SW480 human colon cancer cells." *British Journal of Nutrition* 108 (9): 1623-1632.

Panchal, S. K., Poudyal, H., and Brown, L. 2012. "Quercetin ameliorates cardiovascular, hepatic, and metabolic changes in diet-induced metabolic syndrome in rats." *The Journal of Nutrition* 142 (6): 1026-1032.

Park, H. J., DiNatale, D. A., Chung, M. Y., Park, Y. K., Lee, J. Y., Koo, S. I., O'Connor, M., Manautou, J. E., and Bruno, R. S. 2011. "Green tea extract attenuates hepatic steatosis by decreasing adipose lipogenesis and enhancing hepatic antioxidant defenses in ob/ob mice." *The Journal of Nutritional Biochemistry* 22 (4): 393-400.

Park, S.k., Lee, H. J., Lee, D. H., Lee, S. K., Chun, B. Y., Kim, S. A., Lee, H. S., Son, H. K., and Kim, S.-H. 2008. "Associations of non-alcoholic fatty liver with the metabolic syndrome and serum carotenoids." *Journal of Preventive Medicine and Public Health* 41 (1): 39-44.

Patel, K. R., Andreadi, C., Britton, R. G., Horner-Glister, E., Karmokar, A., Sale, S., Brown, V. A., Brenner, D. E., Singh, R., and Steward, W. P. 2013. "Sulfate metabolites provide an intracellular pool for resveratrol generation and induce autophagy with senescence." *Science Translational Medicine* 5 (205): 205ra133-205ra133.

Paulo, L., Oleastro, M., Gallardo, E., Queiroz, J. A., and Domingues, F. 2011. "Anti-Helicobacter pylori and urease inhibitory activities of resveratrol and red wine." *Food Research International* 44 (4): 964-969.

Pei, Z., Bini, E. J., Yang, L., Zhou, M., Francois, F., and Blaser, M. J. 2004. "Bacterial biota in the human distal esophagus." *Proceedings of the National Academy of Sciences* 101 (12): 4250-4255.

Pei, Z., and Yang, L. 2005. "Bacterial biota in reflux esophagitis and Barrett's esophagus." *World Journal of Gastroenterology* 11 (46): 7277-83.

Peluso, M. R., Winters, T. A., Shanahan, M. F., and Banz, W. J. 2000. "A cooperative interaction between soy protein and its isoflavone-enriched fraction lowers hepatic lipids in male obese Zucker rats and reduces blood platelet sensitivity in male Sprague-Dawley rats." *The Journal of Nutrition* 130 (9): 2333-2342.

Pereg, D., Kotliroff, A., Gadoth, N., Hadary, R., Lishner, M., and Kitay-Cohen, Y. 2011. "Probiotics for patients with compensated liver cirrhosis: a double-blind placebo-controlled study." *Nutrition* 27 (2): 177-181.

Perez-Gregorio, R., and Simal-Gandara, J. 2017. "A critical review of bioactive food components, and of their functional mechanisms, biological effects and health outcomes." *Current Pharmaceutical Design* 23 (19): 2731-2741.

Pérez-Vargas, J. E., Zarco, N., Shibayama, M., Segovia, J., Tsutsumi, V., and Muriel, P. 2014. "Hesperidin Prevents Liver Fibrosis in Rats by Decreasing the Expression of Nuclear Factor-κB, Transforming Growth Factor-ß and Connective Tissue Growth Factor." *Pharmacology* 94 (1-2): 80-89.

Perez-Vizcaino, F., Duarte, J., Jimenez, R., Santos-Buelga, C., and Osuna, A. 2009. "Antihypertensive effects of the flavonoid quercetin." *Pharmacolo-gical Reports* 61 (1): 67-75.

Philpott, H., Nandurkar, S., Lubel, J., and Gibson, P. R. 2015. "Food, fibre, bile acids and the pelvic floor: An integrated low risk low cost approach to managing irritable bowel syndrome." *World Journal of Gastroenterology* 21 (40): 11379-86.

Pinzone, M. R., Celesia, B. M., Di Rosa, M., Cacopardo, B., and Nunnari, G. 2012. "Microbial translocation in chronic liver diseases." *International Journal of Microbiology* 2012: 694629.

Pisonero-Vaquero, S., Martinez-Ferreras, A., Garcia-Mediavilla, M. V., Martinez-Florez, S., Fernandez, A., Benet, M., Olcoz, J. L., Jover, R., Gonzalez-Gallego, J., and Sanchez-Campos, S. 2015. "Quercetin ameliorates dysregulation of lipid metabolism genes via the PI3K/AKT pathway in a diet-induced mouse model of nonalcoholic fatty liver disease." *Molecular Nutrition and Food Research* 59 (5): 879-93.

Poulsen, M. M., Larsen, J. Ø., Hamilton-Dutoit, S., Clasen, B. F., Jessen, N., Paulsen, S. K., Kjær, T. N., Richelsen, B., and Pedersen, S. B. 2012. "Resveratrol up-regulates hepatic uncoupling protein 2 and prevents development of nonalcoholic fatty liver disease in rats fed a high-fat diet." *Nutrition Research* 32 (9): 701-708.

Poulsen, M. M., Vestergaard, P. F., Clasen, B. F., Radko, Y., Christensen, L. P., Stødkilde-Jørgensen, H., Møller, N., Jessen, N., Pedersen, S. B., and Jørgensen, J. O. L. 2013. "High-dose resveratrol supplementation in obese men." *Diabetes* 62 (4): 1186-1195.

Prantera, C., Scribano, M., Falasco, G., Andreoli, A., and Luzi, C. 2002. "Ineffectiveness of probiotics in preventing recurrence after curative resection for Crohn's disease: a randomised controlled trial with Lactobacillus GG." *Gut* 51 (3): 405-409.

Qin, N., Yang, F., Li, A., Prifti, E., Chen, Y., Shao, L., Guo, J., Le Chatelier, E., Yao, J., and Wu, L. 2014. "Alterations of the human gut microbiome in liver cirrhosis." *Nature* 513 (7516): 59-64.

Qiu, L. X., and Chen, T. 2015. "Novel insights into the mechanisms whereby isoflavones protect against fatty liver disease." *World Journal of Gastroenterol* 21 (4): 1099-107.

Quigley, E. M., Stanton, C., and Murphy, E. F. 2013. "The gut microbiota and the liver. Pathophysiological and clinical implications." *Journal of Hepatology* 58 (5): 1020-1027.

Rahimi, R., Abbasabadi, Z., and Abdollahi, M. 2013. "An evidence-based review on medicinal plants used for the treatment of peptic ulcer in traditional Iranian medicine." *International Journal of Pharmacology* 9 (2): 108-124.

Rajan, S., Suganya, H., Thirunalasundari, T., and Jeeva, S. 2012. "Antidiarrhoeal efficacy of Mangifera indica seed kernel on Swiss albino mice." *Asian Pacific Journal of Tropical Medicine* 5 (8): 630-633.

Ramirez-Tortosa, M. C., Ramirez-Tortosa, C. L., Mesa, M. D., Granados, S., Gil, Á., and Quiles, J. L. 2009. "Curcumin ameliorates rabbits's steatohepatitis via respiratory chain, oxidative stress, and TNF-α." *Free Radical Biology and Medicine* 47 (7): 924-931.

Ramkumar, D., and Rao, S. S. 2005. "Efficacy and safety of traditional medical therapies for chronic constipation: systematic review." *The American Journal of Gastroenterology* 100 (4): 936.

Rao, A. V., and Rao, L. G. 2007. "Carotenoids and human health." *Pharmacological Research* 55 (3): 207-216.

Rao, C. V., and Vijayakumar, M. 2008. "Effect of quercetin, flavonoids and alpha-tocopherol, an antioxidant vitamin, on experimental reflux oesophagitis in rats." *European Journal of Pharmacology* 589 (1-3): 233-8.

Rao, S. S. C., Yu, S., and Fedewa, A. 2015. "Systematic review: dietary fibre and FODMAP-restricted diet in the management of constipation and irritable bowel syndrome." *Alimentary Pharmacology & Therapeutics* 41 (12): 1256-1270.

Rayes, N., Seehofer, D., Theruvath, T., Schiller, R. A., Langrehr, J. M., Jonas, S., Bengmark, S., and Neuhaus, P. 2005. "Supply of Pre-and Probiotics Reduces Bacterial Infection Rates After Liver Transplantation—A Randomized, Double-Blind Trial." *American Journal of Transplantation* 5 (1): 125-130.

Rembacken, B., Snelling, A., Hawkey, P., Chalmers, D., and Axon, A. 1999. "Non-pathogenic Escherichia coli versus mesalazine for the treatment of ulcerative colitis: a randomised trial." *The Lancet* 354 (9179): 635-639.

Riggio, O., Varriale, M., Testore, G., Di, R. R., Di, E. R., Merli, M., Romiti, A., Candiani, C., and Capocaccia, L. 1990. "Effect of lactitol and lactulose administration on the fecal flora in cirrhotic patients." *Journal of Clinical Gastroenterology* 12 (4): 433-436.

Ritchie, J. 1973. "Pain from distension of the pelvic colon by inflating a balloon in the irritable colon syndrome." *Gut* 14 (2): 125-132.

Roberfroid, M., and Delzenne, N. M. 1998. "Dietary fructans." *Annual Review of Nutrition* 18 (1): 117-143.

Roberfroid, M. B. 1996. "Functional effects of food components and the gastrointestinal system: chicory fructooligosaccharides." *Nutrition Reviews* 54 (11): 38-42.

Roberfroid, M. B. 2007. "Inulin-type fructans: functional food ingredients." *The Journal of Nutrition* 137 (11): 2493S-2502S.

Rodrigues, F. C., Castro, A. S. B., Rodrigues, V. C., Fernandes, S. A., Fontes, E. A. F., de Oliveira, T. T., Martino, H. S. D., and de Luces Fortes Ferreira, C. L. 2012. "Yacon flour and Bifidobacterium longum modulate bone health in rats." *Journal of Medicinal Food* 15 (7): 664-670.

Rodriguez-Ramiro, I., Vauzour, D., and Minihane, A. 2016. "Polyphenols and non-alcoholic fatty liver disease: impact and mechanisms." *Proceedings of the Nutrition Society* 75 (1): 47-60.

Ryan-Harshman, M., and Aldoori, W. 2004. "How diet and lifestyle affect duodenal ulcers. Review of the evidence." *Canadian Family Physician* 50 (5): 727-732.

Saavedra, J. M., Bauman, N. A., Perman, J., Yolken, R., and Oung, I. 1994. "Feeding of Bifidobacterium bifidum and Streptococcus thermophilus to infants in hospital for prevention of diarrhoea and shedding of rotavirus." *The Lancet* 344 (8929): 1046-1049.

Sadik, R., Abrahamsson, H., Björnsson, E., Gunnarsdottir, A., and Stotzer, P. O. 2003. "Etiology of portal hypertension may influence gastrointestinal transit." *Scandinavian Journal of Gastroenterology* 38 (10): 1039-1044.

Sahebkar, A. 2013. "Why it is necessary to translate curcumin into clinical practice for the prevention and treatment of metabolic syndrome?" *Biofactors* 39 (2): 197-208.

Sahebkar, A. 2014a. "Are curcuminoids effective C-reactive protein-lowering agents in clinical practice? Evidence from a meta-analysis." *Phytotherapy Research* 28 (5): 633-642.

Sahebkar, A. 2014b. "Curcuminoids for the management of hypertriglyce-ridaemia." *Nature Reviews Cardiology* 11 (2): 123-123.

Sahebkar, A. 2014c. "Low-Density Lipoprotein is a Potential Target for Curcumin: Novel Mechanistic Insights." *Basic & Clinical Pharmacology & Toxicology* 114 (6): 437-438.

Sahebkar, A., Chew, G. T., and Watts, G. F. 2014. "Recent advances in pharmacotherapy for hypertriglyceridemia." *Progress in Lipid Research* 56: 47-66.

Sahebkar, A., Mohammadi, A., Atabati, A., Rahiman, S., Tavallaie, S., Iranshahi, M., Akhlaghi, S., Ferns, G. A., and Ghayour-Mobarhan, M. 2013. "Curcuminoids modulate pro-oxidant–antioxidant balance but not the immune response to heat shock protein 27 and oxidized LDL in obese individuals." *Phytotherapy Research* 27 (12): 1883-1888.

Salminen, S., and Salminen, E. 1997. "Lactulose, lactic acid bacteria, intestinal microecology and mucosal protection." *Scandinavian Journal of Gastroenterology* 32 (sup222): 45-48.

Santos, A. M., Lopes, T., Oleastro, M., Gato, I. V., Floch, P., Benejat, L., Chaves, P., Pereira, T., Seixas, E., and Machado, J. 2015. "Curcumin inhibits gastric inflammation induced by Helicobacter pylori infection in a mouse model." *Nutrients* 7 (1): 306-320.

Sasidharan, N. K., Sreekala, S. R., Jacob, J., and Nambisan, B. 2014. "In vitro synergistic effect of curcumin in combination with third generation cephalosporins against bacteria associated with infectious diarrhea." *BioMed Research International* 2014.

Saxelin, M., Tynkkynen, S., Mattila-Sandholm, T., and de Vos, W. M. 2005. "Probiotic and other functional microbes: from markets to mechanisms." *Current Opinion in Biotechnology* 16 (2): 204-211.

Schultz, M., Timmer, A., Herfarth, H. H., Sartor, R. B., Vanderhoof, J. A., and Rath, H. C. 2004. "Lactobacillus GG in inducing and maintaining remission of Crohn's disease." *BMC Gastroenterology* 4 (1): 5.

Seibel, J., Molzberger, A. F., Hertrampf, T., Laudenbach-Leschowski, U., and Diel, P. 2009. "Oral treatment with genistein reduces the expression of molecular and biochemical markers of inflammation in a rat model of chronic TNBS-induced colitis." *European Journal of Nutrition* 48 (4): 213-220.

Seifert, W., Bosma, A., Hendriks, H., Leeuwen, R. v., Ruiter, G., Seifert-Bock, I., Knook, D., and Brouwer, A. 1995. "Beta-carotene (provitamin A) decreases the severity of CCl4-induced hepatic inflammation and fibrosis in rats." *Liver International* 15 (1): 1-8.

Selmi, S., Rtibi, K., Grami, D., Sebai, H., and Marzouki, L. 2017. "Protective effects of orange (Citrus sinensis L.) peel aqueous extract and hesperidin on oxidative stress and peptic ulcer induced by alcohol in rat." *Lipids in Health and Disease* 16 (1): 152.

Setchell, K. D., Brown, N. M., Zimmer-Nechemias, L., Brashear, W. T., Wolfe, B. E., Kirschner, A. S., and Heubi, J. E. 2002. "Evidence for lack of absorption of soy isoflavone glycosides in humans, supporting the crucial role of intestinal metabolism for bioavailability." *The American Journal of Clinical Nutrition* 76 (2): 447-453.

Sgouras, D. N., Panayotopoulou, E. G., Martinez-Gonzalez, B., Petraki, K., Michopoulos, S., and Mentis, A. 2005. "Lactobacillus johnsonii La1 attenuates Helicobacter pylori-associated gastritis and reduces levels of proinflammatory chemokines in C57BL/6 mice." *Clinical and Diagnostic Laboratory Immunology* 12 (12): 1378-1386.

Shadnoush, M., Hosseini, R. S., Mehrabi, Y., Delpisheh, A., Alipoor, E., Faghfoori, Z., Mohammadpour, N., and Moghadam, J. Z. 2013. "Probiotic yogurt affects pro-and anti-inflammatory factors in patients with inflammatory bowel disease." *Iranian Journal of Pharmaceutical Research* 12 (4): 929-936.

Shanahan, F. 2002. "The host–microbe interface within the gut." *Best Practice & Research Clinical Gastroenterology* 16 (6): 915-931.

Shavakhi, A., Minakari, M., Firouzian, H., Assali, R., Hekmatdoost, A., and Ferns, G. 2013. "Effect of a probiotic and metformin on liver aminotransferases in non-alcoholic steatohepatitis: a double blind randomized clinical trial." *International Journal of Preventive Medicine* 4 (5): 531-537.

Shen, J., Zuo, Z.-X., and Mao, A.-P. 2013. "Effect of probiotics on inducing remission and maintaining therapy in ulcerative colitis, Crohn's disease, and pouchitis: meta-analysis of randomized controlled trials." *Inflammatory Bowel Diseases* 20 (1): 21-35.

Shi, T., Fan, G. Q., and Xiao, S. D. 2010. "SIRT3 reduces lipid accumulation via AMPK activation in human hepatic cells." *Journal of Digestive Diseases* 11 (1): 55-62.

Shimoyama, A. T., Santin, J. R., Machado, I. D., e Silva, A. M. d. O., de Melo, I. L. P., Mancini-Filho, J., and Farsky, S. H. 2013. "Antiulcerogenic activity of chlorogenic acid in different models of gastric ulcer." *Naunyn-Schmiedeberg's Archives of Pharmacology* 386 (1): 5-14.

Shin, E. S., Lee, H. H., Cho, S. Y., Park, H. W., Lee, S. J., and Lee, T. R. 2007. "Genistein downregulates SREBP-1 regulated gene expression by inhibiting site-1 protease expression in HepG2 cells." *The Journal of Nutrition* 137 (5): 1127-1131.

Shrestha, S., Ehlers, S. J., Lee, J.Y., Fernandez, M.L., and Koo, S. I. 2009. "Dietary green tea extract lowers plasma and hepatic triglycerides and decreases the expression of sterol regulatory element-binding protein-1c mRNA and its responsive genes in fructose-fed, ovariectomized rats." *The Journal of Nutrition* 139 (4): 640-645.

Simrén, M., Barbara, G., Flint, H. J., Spiegel, B. M., Spiller, R. C., Vanner, S., Verdu, E. F., Whorwell, P. J., and Zoetendal, E. G. 2012. "Intestinal microbiota in functional bowel disorders: a Rome foundation report." *Gut* 62 (1): 159-176.

Singh, P., Singh, N., Sengupta, S., and Palit, G. 2012. "Ameliorative effects of Panax quinquefolium on experimentally induced reflux oesophagitis in rats." *The Indian Journal of Medical Research* 135 (3): 407-413.

Sintara, K., Thong-Ngam, D., Patumraj, S., Klaikeaw, N., and Chatsuwan, T. 2010. "Curcumin suppresses gastric NF-κB activation and macromolecular leakage in Helicobacter pylori-infected rats." *World Journal of Gastroenterology* 16 (32): 4039-4046.

Sivalingam, N., Hanumantharaya, R., Faith, M., Basivireddy, J., Balasubrama-nian, K., and Jacob, M. 2007. "Curcumin reduces indomethacin-induced damage in the rat small intestine." *Journal of Applied Toxicology* 27 (6): 551-560.

Smith, J. J., Kenney, R. D., Gagne, D. J., Frushour, B. P., Ladd, W., Galonek, H. L., Israelian, K., Song, J., Razvadauskaite, G., and Lynch, A. V. 2009. "Small molecule activators of SIRT1 replicate signaling pathways triggered by calorie restriction in vivo." *BMC Systems Biology* 3: 31.

Smith, P. M., Howitt, M. R., Panikov, N., Michaud, M., Gallini, C. A., Bohlooly-y, M., Glickman, J. N., and Garrett, W. S. 2013. "The microbial metabolites, short-chain fatty acids, regulate colonic Treg cell homeostasis." *Science* 341: 6145.

Smith, Y. R., Stohler, C. S., Nichols, T. E., Bueller, J. A., Koeppe, R. A., and Zubieta, J.-K. 2006. "Pronociceptive and antinociceptive effects of estradiol through endogenous opioid neurotransmission in women." *Journal of Neuroscience* 26 (21): 5777-5785.

Snape, W., Shiff, S., and Cohen, S. 1980. "Effect of deoxycholic acid on colonic motility in the rabbit." *American Journal of Physiology-Gastrointestinal and Liver Physiology* 238 (4): G321-G325.

Solmaz, A., Şener, G., Çetinel, Ş., Yüksel, M., Yeğen, C., and Yeğen, B. Ç. 2009. "Protective and therapeutic effects of resveratrol on acetic acid-induced gastric ulcer." *Free Radical Research* 43 (6): 594-603.

Spapen, H., Diltoer, M., Van Malderen, C., Opdenacker, G., Suys, E., and Huyghens, L. 2001. "Soluble fiber reduces the incidence of diarrhea in septic patients receiving total enteral nutrition: a prospective, double-blind, randomized, and controlled trial." *Clinical Nutrition* 20 (4): 301-5.

Spiller, R., Jenkins, D., Thornley, J., Hebden, J., Wright, T., Skinner, M., and Neal, K. 2000. "Increased rectal mucosal enteroendocrine cells, T lymphocytes, and increased gut permeability following acuteCampylobacter enteritis and in post-dysenteric irritable bowel syndrome." *Gut* 47 (6): 804-811.

Spinler, J. K., Taweechotipatr, M., Rognerud, C. L., Ou, C. N., Tumwasorn, S., and Versalovic, J. 2008. "Human-derived probiotic Lactobacillus reuteri demonstrate antimicrobial activities targeting diverse enteric bacterial pathogens." *Anaerobe* 14 (3): 166-171.

Stadlbauer, V., Mookerjee, R. P., Hodges, S., Wright, G. A., Davies, N. A., and Jalan, R. 2008. "Effect of probiotic treatment on deranged neutrophil function and cytokine responses in patients with compensated alcoholic cirrhosis." *Journal of Hepatology* 48 (6): 945-951.

Stahl, W., and Sies, H. 1996. "Lycopene: a biologically important carotenoid for humans?" *Archives of Biochemistry and Biophysics* 336 (1): 1-9.

Stalmach, A., Steiling, H., Williamson, G., and Crozier, A. 2010. "Bioavailability of chlorogenic acids following acute ingestion of coffee by humans with an ileostomy." *Archives of Biochemistry and Biophysics* 501 (1): 98-105.

Steed, H., Macfarlane, G. T., Blackett, K. L., Bahrami, B., Reynolds, N., Walsh, S. V., Cummings, J. H., and Macfarlane, S. 2010. "Clinical trial: the microbiological and immunological effects of synbiotic consumption–a randomized double-blind placebo-controlled study in active Crohn's disease." *Alimentary Pharmacology & Therapeutics* 32 (7): 872-883.

Suerbaum, S., and Michetti, P. 2002. "Helicobacter pylori infection." *New England Journal of Medicine* 347 (15): 1175-1186.

Suskind, D. L., Wahbeh, G., Burpee, T., Cohen, M., Christie, D., and Weber, W. 2013. "Tolerability of curcumin in pediatric inflammatory bowel disease: a forced dose titration study." *Journal of Pediatric Gastroenterology and Nutrition* 56 (3): 277-279.

Suzuki, T. 2013. "Regulation of intestinal epithelial permeability by tight junctions." *Cellular and Molecular Life Sciences* 70 (4): 631-659.

Suzuki, T., and Hara, H. 2011. "Role of flavonoids in intestinal tight junction regulation." *The Journal of Nutritional Biochemistry* 22 (5): 401-408.

Swarnakar, S., Ganguly, K., Kundu, P., Banerjee, A., Maity, P., and Sharma, A. V. 2005. "Curcumin regulates expression and activity of matrix metalloproteinases 9 and 2 during prevention and healing of indomethacin-induced gastric ulcer." *Journal of Biological Chemistry* 280 (10): 9409-9415.

Swidsinski, A., Ladhoff, A., Pernthaler, A., Swidsinski, S., Loening–Baucke, V., Ortner, M., Weber, J., Hoffmann, U., Schreiber, S., and Dietel, M. 2002. "Mucosal flora in inflammatory bowel disease." *Gastroenterology* 122 (1): 44-54.

Szkudelski, T., and Szkudelska, K. 2011. "Anti-diabetic effects of resveratrol." *Annals of the New York Academy of Sciences* 1215 (1): 34-39.

Tadesse, E., Engidawork, E., Nedi, T., and Mengistu, G. 2017. "Evaluation of the anti-diarrheal activity of the aqueous stem extract of Lantana camara Linn (Verbenaceae) in mice." *BMC Complementary and Alternative Medicine* 17 (1): 190.

Tainaka, T., Shimada, Y., Kuroyanagi, J., Zang, L., Oka, T., Nishimura, Y., Nishimura, N., and Tanaka, T. 2011. "Transcriptome analysis of anti-fatty liver action by Campari tomato using a zebrafish diet-induced obesity model." *Nutrition & Metabolism* 8 (1): 88.

Tandon, R., Khanna, R., Dorababu, M., and Goel, R. 2004. "Oxidative stress and antioxidants status in peptic ulcer and gastric carcinoma." *Indian Journal of Physiology and Pharmacology* 48 (1): 115-118.

Tannock, G. W., Wilson, C. M., Loach, D., Cook, G. M., Eason, J., O'toole, P. W., Holtrop, G., and Lawley, B. 2012. "Resource partitioning in relation to cohabitation of Lactobacillus species in the mouse forestomach." *The ISME journal* 6 (5): 927-938.

Tarpila, A., Wennberg, T., and Tarpila, S. 2005. "Flaxseed as a functional food." *Current Topics in Nutraceutical Research* 3 (3): 167.

Thompson, W., Heaton, K., Smyth, G., and Smyth, C. 2000. "Irritable bowel syndrome in general practice: prevalence, characteristics, and referral." *Gut* 46 (1): 78-82.

Thompson, W. G., Longstreth, G., Drossman, D., Heaton, K., Irvine, E., and Müller-Lissner, S. 1999. "Functional bowel disorders and functional abdominal pain." *Gut* 45 (suppl 2): II43-II47.

Tomé-Carneiro, J., Larrosa, M., González-Sarrías, A., A Tomas-Barberan, F., Teresa Garcia-Conesa, M., and Carlos Espin, J. 2013. "Resveratrol and clinical trials: the crossroad from in vitro studies to human evidence." *Current Pharmaceutical Design* 19 (34): 6064-6093.

Trowell, H. C. 1975. *Refined Carbohydrate Foods and Disease: Some Implications of Dietary Fibre*. Academic.

Turner, J. R. 2009. "Intestinal mucosal barrier function in health and disease." *Nature Reviews Immunology* 9 (11): 799.

Um, M. Y., Hwang, K. H., Ahn, J., and Ha, T. Y. 2013. "Curcumin Attenuates Diet-Induced Hepatic Steatosis by Activating AMP-Activated Protein Kinase." *Basic & Clinical Pharmacology & Toxicology* 113 (3): 152-157.

Unlu, N. Z., Bohn, T., Clinton, S. K., and Schwartz, S. J. 2005. "Carotenoid absorption from salad and salsa by humans is enhanced by the addition of avocado or avocado oil." *The Journal of Nutrition* 135 (3): 431-436.

Usami, M., Miyoshi, M., Kanbara, Y., Aoyama, M., Sakaki, H., Shuno, K., Hirata, K., Takahashi, M., Ueno, K., and Hamada, Y. 2013. "Analysis of fecal microbiota, organic acids and plasma lipids in hepatic cancer patients with or without liver cirrhosis." *Clinical Nutrition* 32 (3): 444-451.

Usami, M., Miyoshi, M., Kanbara, Y., Aoyama, M., Sakaki, H., Shuno, K., Hirata, K., Takahashi, M., Ueno, K., Tabata, S., Asahara, T., and Nomoto, K. 2011. "Effects of perioperative synbiotic treatment on infectious complications, intestinal integrity, and fecal flora and organic acids in hepatic surgery with or without cirrhosis." *JPEN. Journal of Parenteral and Enteral Nutrition* 35 (3): 317-28.

Vajro, P., Mandato, C., Licenziati, M. R., Franzese, A., Vitale, D. F., Lenta, S., Caropreso, M., Vallone, G., and Meli, R. 2011. "Effects of Lactobacillus rhamnosus strain GG in pediatric obesity-related liver disease." *Journal of Pediatric Gastroenterology and Nutrition* 52 (6): 740-743.

Vanderhoof, J., Murray, N. D., Paule, C. L., and Ostrom, K. M. 1997. "Use of soy fiber in acute diarrhea in infants and toddlers." *Clinical Pediatrics* 36 (3): 135-139.

Vanni, E., and Bugianesi, E. 2009. "The gut-liver axis in nonalcoholic fatty liver disease: Another pathway to insulin resistance?" *Hepatology* 49 (6): 1790-1792.

Vasiljevic, T., and Shah, N. P. 2008. "Probiotics—from Metchnikoff to bioactives." *International Dairy Journal* 18 (7): 714-728.

Velázquez, C., Correa-Basurto, J., Garcia-Hernandez, N., Barbosa, E., Tesoro-Cruz, E., Calzada, S., and Calzada, F. 2012. "Anti-diarrheal activity of (–)-Epicatechin from Chiranthodendron pentadactylon Larreat: Experimental and computational studies." *Journal of Ethnopharmacology* 143 (2): 716-719.

Verdú, E. F., Deng, Y., Bercik, P., and Collins, S. M. 2002. "Modulatory effects of estrogen in two murine models of experimental colitis." *American Journal of Physiology-Gastrointestinal and Liver Physiology* 283 (1): G27-G36.

Vidyashankar, S., Varma, R. S., and Patki, P. S. 2013. "Quercetin ameliorate insulin resistance and up-regulates cellular antioxidants during oleic acid induced hepatic steatosis in HepG2 cells." *Toxicology in Vitro* 27 (2): 945-953.

Vos, M. B. 2014. "Nutrition, nonalcoholic fatty liver disease and the microbiome: recent progress in the field." *Current Opinion in Lipidology* 25 (1): 61.

Wacklin, P., Kaukinen, K., Tuovinen, E., Collin, P., Lindfors, K., Partanen, J., Mäki, M., and Mättö, J. 2013. "The duodenal microbiota composition of adult celiac disease patients is associated with the clinical manifestation of the disease." *Inflammatory Bowel Diseases* 19 (5): 934-941.

Wacklin, P., Laurikka, P., Lindfors, K., Collin, P., Salmi, T., Lahdeaho, M. L., Saavalainen, P., Maki, M., Matto, J., Kurppa, K., and Kaukinen, K. 2014. "Altered duodenal microbiota composition in celiac disease patients suffering from persistent symptoms on a long-term gluten-free diet." *American Journal of Gastroenterology* 109 (12): 1933-41.

Wada-Hiraike, O., Imamov, O., Hiraike, H., Hultenby, K., Schwend, T., Omoto, Y., Warner, M., and Gustafsson, J.-Å. 2006. "Role of estrogen receptor β in colonic epithelium." *Proceedings of the National Academy of Sciences of the United States of America* 103 (8): 2959-2964.

Wang, G.-L., Fu, Y.-C., Xu, W.-C., Feng, Y.-Q., Fang, S.-R., and Zhou, X.-H. 2009. "Resveratrol inhibits the expression of SREBP1 in cell model of steatosis via Sirt1–FOXO1 signaling pathway." *Biochemical and Biophysical Research Communications* 380 (3): 644-649.

Wang, H.-B., Wang, P.-Y., Wang, X., Wan, Y.-L., and Liu, Y.-C. 2012. "Butyrate enhances intestinal epithelial barrier function via up-regulation of tight junction protein Claudin-1 transcription." *Digestive Diseases and Sciences* 57 (12): 3126-3135.

Wang, L.-S., Kuo, C.-T., Huang, T. H.-M., Yearsley, M., Oshima, K., Stoner, G. D., Yu, J., Lechner, J. F., and Huang, Y.-W. 2013. "Black raspberries protectively regulate methylation of Wnt pathway genes in precancerous colon tissue." *Cancer Prevention Research* 6 (12): 1317-1327.

Wang, M.-E., Chen, Y.-C., Chen, I.-S., Hsieh, S.-C., Chen, S.-S., and Chiu, C.-H. 2012. "Curcumin protects against thioacetamide-induced hepatic fibrosis by attenuating the inflammatory response and inducing apoptosis of damaged hepatocytes." *The Journal of Nutritional Biochemistry* 23 (10): 1352-1366.

Wang, X., Hasegawa, J., Kitamura, Y., Wang, Z., Matsuda, A., Shinoda, W., Miura, N., and Kimura, K. 2011. "Effects of hesperidin on the progression of hypercholesterolemia and fatty liver induced by high-cholesterol diet in rats." *Journal of Pharmacological Sciences* 117 (3): 129-138.

Wang, Y.-H., Ge, B., Yang, X.-L., Zhai, J., Yang, L.-N., Wang, X.-X., Liu, X., Shi, J.-C., and Wu, Y.-J. 2011. "Proanthocyanidins from grape seeds modulates the nuclear factor-kappa B signal transduction pathways in rats with TNBS-induced recurrent ulcerative colitis." *International Immunopharmacology* 11 (10): 1620-1627.

Wardi, J., Reifen, R., Aeed, H., Zadel, L., Avni, Y., and Bruck, R. 2001. "Beta-carotene attenuates experimentally induced liver cirrhosis in rats." *The Israel Medical Association Journal: IMAJ* 3 (2): 151-154.

Weill, P., Schmitt, B., Chesneau, G., Daniel, N., Safraou, F., and Legrand, P. 2002. "Effects of introducing linseed in livestock diet on blood fatty acid composition of consumers of animal products." *Annals of Nutrition and Metabolism* 46 (5): 182-191.

Wergedahl, H., Liaset, B., Gudbrandsen, O. A., Lied, E., Espe, M., Muna, Z., Mørk, S., and Berge, R. K. 2004. "Fish protein hydrolysate reduces plasma total cholesterol, increases the proportion of HDL cholesterol, and lowers acyl-CoA: cholesterol acyltransferase activity in liver of Zucker rats." *The Journal of Nutrition* 134 (6): 1320-1327.

Wiest, R., Lawson, M., and Geuking, M. 2014. "Pathological bacterial translocation in liver cirrhosis." *Journal of Hepatology* 60 (1): 197-209.

Wong, V., Won, G., Chim, A., Chu, W., Yeung, D., Li, K., and Chan, H. 2013. "Treatment of nonalcoholic steatohepatitis with probiotics. A proof-of-concept study." *Annals of Hepatology* 12 (2): 256-262.

Wongsamitkul, N., Sirianant, L., Muanprasat, C., and Chatsudthipong, V. 2010. "A plant-derived hydrolysable tannin inhibits CFTR chloride channel: a potential treatment of diarrhea." *Pharmaceutical Research* 27 (3): 490-497.

World Health Organization and Food & Agriculture Organization, F. j. 2002. "WHO working group report on drafting guidelines for the evaluation of probiotics in food." *London, Ontario, Canada* 30.

Xin, P., Han, H., Gao, D., Cui, W., Yang, X., Ying, C., Sun, X., and Hao, L. 2013. "Alleviative effects of resveratrol on nonalcoholic fatty liver disease are associated with up regulation of hepatic low density lipoprotein receptor and scavenger receptor class B type I gene expressions in rats." *Food and Chemical Toxicology* 52: 12-18.

Xu, H. E., Lambert, M. H., Montana, V. G., Parks, D. J., Blanchard, S. G., Brown, P. J., Sternbach, D. D., Lehmann, J. M., Wisely, G. B., and Willson, T. M. 1999. "Molecular recognition of fatty acids by peroxisome proliferator–activated receptors." *Molecular Cell* 3 (3): 397-403.

Xu, J., Gao, H., Song, L., Yang, W., Chen, C., Deng, Q., Huang, Q., and Huang, F. 2013. "Flaxseed oil and alpha-lipoic acid combination ameliorates hepatic oxidative stress and lipid accumulation in comparison to lard." *Lipids in Health and Disease* 12 (1): 58.

Xu, J., Ma, R., Chen, L. F., Zhao, L. J., Chen, K., and Zhang, R. B. 2014. "Effects of probiotic therapy on hepatic encephalopathy in patients with liver cirrhosis: an updated meta-analysis of six randomized controlled trials." *Hepatobiliary and Pancreatic Diseases International* 13 (4): 354-60.

Xu, J., Zhou, X., Chen, C., Deng, Q., Huang, Q., Yang, J. e., Yang, N., and Huang, F. 2012. "Laxative effects of partially defatted flaxseed meal on normal and experimental constipated mice." *BMC Complementary and Alternative Medicine* 12: 14.

Yalniz, M., Bahcecioglu, I. H., Kuzu, N., Poyrazoglu, O. K., Bulmus, O., Celebi, S., Ustundag, B., Ozercan, I. H., and Sahin, K. 2007. "Preventive role of genistein in an experimental non-alcoholic steatohepatitis model." *Journal of Gastroenterology and Hepatology* 22 (11): 2009-2014.

Yang, L., Chaudhary, N., Baghdadi, J., and Pei, Z. 2014. "Microbiome in reflux disorders and esophageal adenocarcinoma." *The Cancer Journal* 20 (3): 207-10.

Yang, L., Francois, F., and Pei, Z. 2012. "Molecular pathways: pathogenesis and clinical implications of microbiome alteration in esophagitis and Barrett esophagus." *Clinical Cancer Research* 18 (8): 2138-44.

Yang, Y.-J., Chuang, C.-C., Yang, H.-B., Lu, C.-C., and Sheu, B.-S. 2012. "Lactobacillus acidophilus ameliorates H. pylori-induced gastric inflammation by inactivating the Smad7 and NFκB pathways." *BMC Microbiology* 12 (1): 38.

Yari, Z., Rahimlou, M., Eslamparast, T., Ebrahimi-Daryani, N., Poustchi, H., and Hekmatdoost, A. 2016. "Flaxseed supplementation in non-alcoholic fatty liver disease: a pilot randomized, open labeled, controlled study." *International Journal of Food Sciences and Nutrition* 67 (4): 461-469.

Ying, H.-Z., Liu, Y.-H., Yu, B., Wang, Z.-Y., Zang, J.-N., and Yu, C.-H. 2013. "Dietary quercetin ameliorates nonalcoholic steatohepatitis induced by a high-fat diet in gerbils." *Food and Chemical Toxicology* 52: 53-60.

Young, D., Ibuki, M., Nakamori, T., Fan, M., and Mine, Y. 2011. "Soy-Derived Di-and Tripeptides Alleviate Colon and Ileum Inflammation in Pigs with Dextran Sodium Sulfate-Induced Colitis–3." *The Journal of Nutrition* 142 (2): 363-368.

Zaidi, S. F. H., Yamamoto, T., Refaat, A., Ahmed, K., Sakurai, H., Saiki, I., Kondo, T., Usmanghani, K., Kadowaki, M., and Sugiyama, T. 2009. "Modulation of Activation-Induced Cytidine Deaminase by Curcumin in Helicobacter Pylori-Infected Gastric Epithelial Cells." *Helicobacter* 14 (6): 588-595.

Zang, M., Xu, S., Maitland-Toolan, K. A., Zuccollo, A., Hou, X., Jiang, B., Wierzbicki, M., Verbeuren, T. J., and Cohen, R. A. 2006. "Polyphenols stimulate AMP-activated protein kinase, lower lipids, and inhibit accelerated atherosclerosis in diabetic LDL receptor–deficient mice." *Diabetes* 55 (8): 2180-2191.

Zhang, C., Yuan, W., Fang, J., Wang, W., He, P., Lei, J., and Wang, C. 2016. "Efficacy of Resveratrol Supplementation against Non-Alcoholic Fatty Liver Disease: A Meta-Analysis of Placebo-Controlled Clinical Trials." *PLoS One* 11 (8): e0161792.

Zhang, D., Xie, L., Jia, G., Cai, S., Ji, B., Liu, Y., Wu, W., Zhou, F., Wang, A., and Chu, L. 2011. "Comparative study on antioxidant capacity of flavonoids and their inhibitory effects on oleic acid-induced hepatic steatosis in vitro." *European Journal of Medicinal Chemistry* 46 (9): 4548-4558.

Zhang, T., DeSimone, R. A., Jiao, X., Rohlf, F. J., Zhu, W., Gong, Q. Q., Hunt, S. R., Dassopoulos, T., Newberry, R. D., and Sodergren, E. 2012. "Host genes related to paneth cells and xenobiotic metabolism are associated with shifts in human ileum-associated microbial composition." *PloS one* 7 (6): e30044.

Zhang, Y., Li, L., Guo, C., Mu, D., Feng, B., Zuo, X., and Li, Y. 2016. "Effects of probiotic type, dose and treatment duration on irritable bowel syndrome diagnosed by Rome III criteria: a meta-analysis." *BMC Gastroenterology* 16 (1): 62.

Zhao, N. J., Liao, M. J., Wu, J. J., and Chu, K. X. 2018. "Curcumin suppresses Notch1 signaling: Improvements in fatty liver and insulin resistance in rats." *Molecular Medicine Reports* 17 (1): 819-826.

Zhao, R., Wang, Y., Huang, Y., Cui, Y., Xia, L., Rao, Z., Zhou, Y., and Wu, X. 2017. "Effects of fiber and probiotics on diarrhea associated with enteral nutrition in gastric cancer patients: A prospective randomized and controlled trial." *Medicine (Baltimore)* 96 (43): e8418.

Zhao, Y., and Yu, Y.-B. 2016. "Intestinal microbiota and chronic constipation." *SpringerPlus* 5 (1): 1130.

Zhou, Q., Zhang, B., and Verne, G. N. 2009. "Intestinal membrane permeability and hypersensitivity in the irritable bowel syndrome." *Pain* 146 (1-2): 41-46.

Zhu, L., Liu, W., Alkhouri, R., Baker, R. D., Bard, J. E., Quigley, E. M., and Baker, S. S. 2014. "Structural changes in the gut microbiome of constipated patients." *Physiological Genomics* 46 (18): 679-686.

Zong, G., Demark-Wahnefried, W., Wu, H., and Lin, X. 2013. "Effects of flaxseed supplementation on erythrocyte fatty acids and multiple cardiometabolic biomarkers among Chinese with risk factors of metabolic syndrome." *European Journal of Nutrition* 52 (5): 1547-1551.

Zong, H., Ren, J. M., Young, L. H., Pypaert, M., Mu, J., Birnbaum, M. J., and Shulman, G. I. 2002. "AMP kinase is required for mitochondrial biogenesis in skeletal muscle in response to chronic energy deprivation." *Proceedings of the National Academy of Sciences* 99 (25): 15983-15987.

About the Authors

Amir M. Mortazavian

Amir M. Mortazavian graduated from University of Tehran, Iran (BSc, MSc and PhD of Food Technology). He is from the top 1% ESI (Essential Science Indicators) and has experience in teaching, research (numerous research awards and honors), executive work and the food industry. He has published several international books and chapters of books as well as more than 200 international articles indexed in Scopus and ISI Web of Science. His professional interests are probiotics in food, functional foods, bioactive compounds in food, dairy science, and the chemistry of food matrices. He is a professor within the faculty of Nutrition Sciences and Food Technology, National Nutrition and Food Technology Research Institute (Iran).

Zahra Yari

Zahra Yari graduated from Shahid Beheshti University of Medical Sciences, Iran (BSc, MSc and PhD of Nutrition and Dietetics). She has published several book chapters and international papers indexed in Scopus and ISI Web of Science. She has been working in the field of nutrition and clinical diet therapy for almost 10 years and has high skills in clinical nutrition. Her professional interests are mainly nutrition and diet therapy in gastrointestinal, endocrine, cardiovascular and kidney diseases and the use of functional foods in the treatment of these diseases.

Nasim Khorshidian

Nasim Khorshidian graduated from Shahid Beheshti University of Medical Sciences, Iran (BSc, MSc and PhD of Food Technology). She is an assistant professor within the Department of Food Technology Research at the National Nutrition and Food Technology Research Institute, Shahid Beheshti University of Medical Sciences. She has published several book chapters and international papers indexed in Scopus and ISI Web of Science. Her professions and interests are probiotics in food, functional foods, food chemistry and food safety.

Index

B

C

D

E

F

G

H

I

N

O

P